Jon Attwood

Series editor: *Geoff Hancock*

Graphic Products

Edexcel A Level Design and Technology:
Product Design

A PEARSON COMPANY

British Library Cataloguing in Publication Data
A catalogue record for this book is available from the British Library.

ISBN 978 0 435757 79 3

Edited by Stephen Nicholls
Designed by Wooden Ark Studios
Typeset by Tek-Art
Original illustrations © Pearson Education Ltd
Cover design by Wooden Ark Studios
Picture research by Zooid Pictures (Ned Coombes)
Cover photo/illustration © Allaamy / Adrian Sherratt
Printed in China (GCC/06)

Acknowledgements
The author and publisher would like to thank the following individuals and organisations for permission to reproduce photographs:
Unit 2: Alamy / Rob Wilkinson **pg 34**; Alamy / PCL **pg 35**; Pearson Education Ltd / Trevor Clifford **pg 36**; Alamy / Ferruccio **pg 38**; Alamy / Eddie Gerald **pg 39** Corbis / Pete Leonard / Zefa **pg 40** Alamy / Ian Nolan **pg 40** Alamy / Chris George **pg 43** Alamy / Paul Whitehill **pg 43** Zooid Pictures / Ned Coomes; **pg 44**; Fujitsu **pg 44**; Science Photo Library **pg 45**; Alamy / kolvenbach **pg 45**; Nike UK Ltd **pg 47**; Corbis / Radim Beznoska / Epa **pg 48**; Dyson; **pg 54** Zooid Pictures / Ned Coomes; **pg 63**; Celloglas **pg 63**; Inspeck **pg 64**; BMW **pg 65**; Science Photo Library **pg 71**; Techproducts **pg 71**; BSI **pg 73**; Pearson Education Ltd / Tudor Photography **pg 73**; Evoluent **pg 77**; *Unit 3:* Getty Images / Dan Krauss **pg 87**; Getty Images / Louie Psihoyos / Science Faction **pg 89**; Science Photo Library **pg 89**; Getty Images / Jeff Smith / The; Image Bank; **pg 89**; Science Photo Library **pg 91**; Getty Images / Michael Courtney / Warner Bros. **pg 94**; Getty Images / Yoshikazu Tsuno / AFP **pg 95**; Innocent Drinks **pg 98**; Science Photo Library **pg 107**; Science Photo Library **pg 109**; Alamy / Corbis Premium RF **pg 109**; Getty Images / Spencer Platt **pg 110**; Getty Images / H. Armstrong Roberts / Retrofile **pg 113**; Getty Images / Bill Kalis / Time & Life Pictures **pg 114**; Nokia **pg 116**; Audi UK **pg 117**; akg-images **pg 120**; Alamy / David Ball **pg 121**; Bridgeman Art Library **pg 121**; Alamy / eddie linssen **pg 122**; Corbis / Peter Harholdt **pg 123**; Bridgeman Art Library **pg 123**; Zooid Pictures **pg 124**; Alamy / Yadid Levy **pg 124**; Alamy / imagebroker **pg 125**; akg-images **pg 126**; Alamy / dk **pg 127**; Erno Goldfinger / www.aaschool.ac.uk **pg 128**; Corbis / Fernando Bengoechea / Beateworks **pg 128**; Harry Ransom Humanities Research Center The University of Texas at Austin **pg 129**; Volkswagen Head Office UK **pg 129**; Corbis / Bettmann **pg 130**; Getty Images / Arnaud Chicurel / hemis.fr **pg 131**; Carlton, Memphis 1981 design by Ettore Sottsass / Thomas Dix / Vitra Design; Museum **pg 132**; Rex Features / Stephane **pg 133**; Alamy / Redfx **pg 133**; House for an Art Lover **pg 134**; Alessi / Hot Bertaa design Philippe Starck for Alessi **pg 136**; iStockPhoto.com / Alexandra Draghici **pg 136**; Alamy / Chris Willson **pg 139**; Alamy / Motoring Picture Library **pg 145**; Sennheiser **pg 146**; Alamy /Chris Laurens **pg 155**; 1996 Forest Stewardship Council A.C. **pg 157**; Alamy / Richard Sheppard **pg 140**; Alamy / Redfx **pg 162**; Alamy / Flora Press **pg 162**; Shutterstock / Franck Boston *Unit 4:* **pg 183**; iStockPhoto.com **pg 184**; iStockPhoto.com / Luis Carlos Torres **pg 185.**

Author supplied images: pg 4, pg 6, pg 8, pg 10, pg 12, pg 13, pg 14, pg 15, pg 16, pg 17, pg 19, pg 20, pg 23, pg 25, pg 27, pg 164, pg 165, pg 169, pg 170, pg 172, pg 174, pg 176 and pg 182.

Websites
There are links to relevant websites in this book. In order to ensure that these are up to date, that they work, and that the sites are not inadvertently linked to sites that could be considered offensive, we have made the links available on the Heinemann website at www.heinemann.co.uk/hotlinks. When you access the site, the express code is 7793p.

Contents

Introduction

This book is designed to support Edexcel's GCE Graphic Products specification. The book directly follows the structure of the specification and seeks to develop your knowledge, understanding, skills and application for designing products. Graphic Products encompasses a wide range of design disciplines but is firmly rooted in the skills required to design and make high-quality products – products that are fit for purpose, satisfy wants and needs, enhance our day-to-day lives and, most importantly, give you the opportunity to demonstrate your design and technology capability.

This qualification emphasises two key factors: creativity and sustainability. We want you to explore ideas of originality and value, to question and challenge, to envisage what could be, but equally to achieve the results that will progress your career. This new qualification structure allows you to develop a range of skills and outcomes at AS level which will demonstrate your creativity and you will be able to apply these to a design-and-make project at A2.

All modern designers have to consider sustainable issues when designing new products. A sign of the modern technological age in which we live is that human actions have had a negative impact on our environment. New products should provide solutions rather than add to the existing problems of extractions and use of natural resources, pollution from manufacturing and disposal of large amounts of waste products.

Good design is vital to our world and economy; it is important, therefore, that you, as a future designer, develop a passion for designing your future.

How to use this book

This book is divided into four parts which correspond to the four units of the course: AS Units 1 and 2, and A2 Units 3 and 4. These sections will provide you with a depth of knowledge and understanding of Graphic Products to help you study for examined units and to help you better understand the requirements of coursework units. All units follow the headings stated in the Edexcel specification content so that all the relevant information is covered in detail.

Table i *Structure of the book's AS/A2 units*

Summary of expectations for the unit	Unit content	Exam café
The first section of the unit summarises: • what you are required to do • what you will learn • how the unit is assessed.	This main section covers the subject content in depth. It: • explains what you will learn in each unit • helps you to understand the assessment requirements • provides further support and guidance throughout the unit.	The last section of examined units contains a revision section comprising: • a teacher area involving tips for studying • an 'ask the examiner' section involving worked examples of exam questions and practice exam questions.

Support and guidance

A number of headings and boxes feature throughout the book designed to give you further support and guidance where appropriate.

General support and guidance common to all units.

• **'Getting Started!'** introduces each new section and asks some key questions relating to the content of each topic.

These help you to start thinking about how much you really know about a topic before you study it in depth.

- **'Factfile'** boxes contain information which may explain technical terms or further illustrate points in the main text.
- **'Weblinks'** provide suggestions for further study of key topics by giving relevant websites for you to research.
- **'Links to'** show you how information in one section can be cross-referenced to another section to help you fully understand a topic.

Support and guidance specific to examined Units 2 and 3.

- **'Think about this'** sections pose questions relating to the topic just covered. These are designed to test your knowledge and understanding of the topic you have just studied.
- There are a significant number of tables throughout examined units which will help you to revise significant points relating to many topics. In addition, the widespread use of figures helps to illustrate the main text.

Support and guidance specific to coursework Units 1 and 4.

- **'Activity'** boxes suggest relevant tasks you may want to undertake in order to develop your design and technology capability.
- **'To be successful you will'** boxes appear at the end of each coursework section enabling you to understand the assessment criteria and what you will need to evidence in order to gain as many marks as you can.
- Students' work is illustrated throughout coursework units to give you an idea of both the layout and content of each assessment criteria. They are examples that show good practice and should not be simply copied.

Exam café

An **'Exam café'** section features at the end of examined Units 2 and 3. This is an extremely useful section to use when revising for your exams and comprises two main sections.

1. Teacher area.
- A **'Revision summary'** gives you important information about studying for your exams.

- A **'Revision checklist'** provides you with a useful list of things to do when revising.
- **'Tips for answering questions'** gives you an invaluable guide to the command words used in each exam so you know what to expect from each question.

2. Ask the examiner.
- **'Worked examples'** for four questions are provided for each unit which show you what the examiner will be looking for in a range of different styles of question. The examiner will guide you through what each question is actually asking you to do and show you where marks will be awarded. High and low responses are used to illustrate each question.
- A number of **'Practice questions'** are given which directly relate to the content of each unit. These will prove extremely useful in testing your knowledge and understanding of a topic and provide invaluable exam practice.

How this course is structured

This is a four-unit course: two units at AS level and a further two units at A2 level which combine to make the full GCE qualification. The course will be assessed by both externally set examinations and internal assessment. For further guidance and a full outline of the content of Units 1–4 please refer to the **'Summary of expectations'** at the start of each unit in this book.

AS units

Unit 1: Portfolio of Creative Skills	Unit 2: Design and Technology in Practice
Internal assessment Internally set and marked by your school/college and externally moderated by Edexcel. **Number of marks: 90**	**External assessment** Time: 1-hour 30-minute examination set and marked by Edexcel. **Number of marks: 70**
You produce one portfolio that contains evidence of product investigation, product design and product manufacture. Photographic evidence must be supplied for the product(s) you have made.	You complete a question and answer booklet, consisting of short-answer and extended-writing type questions.
60% of AS course 30% of full GCE	**40% of AS course 20% of full GCE**

A2 units

Unit 3: Designing for the Future	Unit 4: Commercial Design
External assessment Time: 2-hour examination set and marked by Edexcel. Number of marks: 70	Internal assessment Internally set and marked by your school/college and externally moderated by Edexcel. Number of marks: 90
You complete a question and answer booklet, consisting of short-answer and extended-writing type questions.	You design and make a product. This is evidenced in your design folder with photographic evidence of you making the product and of the final product itself.
40% of A2 course 20% of full GCE	60% of A2 course 30% of full GCE

General exam advice

Preparation

'If you fail to prepare – prepare to fail' is probably the best piece of advice given to any student when it comes to taking exams. Therefore, it is important that you have developed a thorough knowledge and understanding of the topics covered and not just 'crammed' a few days before the exam. Useful preparation can include the following.

- **Organise your notes** as an organised file; this will save you hours when the time comes to revise. You could create:
  - Summary notes and "mind-maps" which outline the most important ideas of each topic in a visual manner instead of blocks of text.
  - Flashcards containing information that you need to have memorised. You could put topics on one side of the card, answers on the other. Flashcards will enable you to test your ability to not only recognise important information, but also your ability to retrieve information from memory.
  - Study checklists that identify the topics that you will be tested on for each exam. This checklist will enable you to break your studying into organised, manageable chunks, which should allow for a comprehensive revision plan with minimal anxiety.

- **Frequently read back over your notes after a lesson** to make sure that you understand the topic. Simply copying down notes from the board or from the textbook, without giving much thought to the content of your notes, will not be useful when you come to revise.

- **Frequently review your notes before the next lesson** and highlight any questions you may have. Do not be afraid of asking your teacher questions if you do not understand a topic.

- **Frequently test yourself on each topic** using the practice exam questions in the Exam café sections of this book and the 'Think about this' guidance throughout each unit.

- **Start to plan your revision early** so that you have more than enough time to cover all the topics in this exam and exams in your other subjects.

Revision and practice

Revision and practice are crucial to exam success. An exam not only tests your knowledge and understanding of topics but asks you to apply this to various contexts within each question. In order to build your confidence you should aim to:

- **Familiarise yourself with the format of the exam** using sample assessment materials and past papers so that there are no surprises on the actual exam day. Know how long you have and what sort of questions have been asked in the past.

- **Practise answering past questions** which provide you with invaluable experience of completing answers in a given time. Using exam conditions helps improve your planning and writing skills in producing focused responses.

- **Make the most of trial exams** as they help you to identify your strengths and weaknesses. Continue to revise topics that you are familiar with but do not avoid improving areas of weakness.

On the day

Everybody suffers from exam anxiety, which causes stress. However, there are some simple things that you can try which can manage your anxiety and reduce your stress levels.

- Prepare thoroughly and learn the topics well. Approach the exam with confidence and view it as an opportunity to show how much you have studied.

- Get a good night's sleep the night before and make sure you don't go to the exam on an empty stomach.
- Allow yourself plenty of time to get to school/college and even plan to get there a little early so you have time to relax before the exam.
- Don't attempt last minute cramming – you should feel confident that you have prepared.

During the exam

Now that you are actually in the exam hall and ready to start the exam there are a number of tips that can help you to perform to the best of your ability.

- **Preview the entire exam paper** by spending a short period reading through the paper carefully, marking key terms and deciding how to budget your time. Plan to do the easy questions first and the most difficult questions last.
- **Read each question carefully**, making sure that you know what the question is actually asking you and what is required in your response.
- **Plan your responses**, especially to extended writing type questions or essays. Use a blank piece of paper to jot down the key points and structure.
- **Write in a clear and legible manner** using black or blue ink so that the examiner can read your responses.
- **If you go blank then move onto the next question** and don't start getting stressed about the last one. You should have time to go back to that difficult question at the end of the exam.
- **Check your answers** and don't simply sit there staring into space if you have time to spare. You may spot a mistake or want to add more detail to a particular part – if you don't know the answer then have an 'educated guess'.

Examiners

All Edexcel examiners want you to do your best. None of them want to try to catch you out or fail you; indeed, examiners are sometimes briefed to give 'benefit of the doubt' to responses that are nearly correct. All examiners work from a mark scheme which is written by the principal examiner (who wrote the exam) and discussed in detail with examiners before they begin marking. A mark scheme indicates several acceptable responses to each question and also what is not acceptable. Therefore, examiners will be looking for 'triggers' in your work which relate directly back to their mark scheme.

It is extremely important to **read each question** thoroughly before you answer it. It is common for examiners to complain that too many students fail to answer the question that has actually been asked. Sometimes students try to write everything they know about the topic being examined instead of writing a focused response which can be allocated marks. It is also a common mistake to read the question too quickly and misunderstand what's being asked. For example, being asked for the *disadvantages* of a process rather than its *advantages*. It is often a very useful exercise to highlight the key words before you answer to make sure that you are giving the examiner the correct response.

Glossary of Terms

It would be extremely useful for you to have a copy of the specific unit content for the examined Units 2 and 3. These can be photocopied from your teacher's copy of the Edexcel specification or downloaded at www.edexcel.org.uk. Each section in the unit content carries a 'stem' explaining what you specifically need to learn for each examination.

Unit 2: Design and Technology in Practice

For example:
b) Polymers

Aesthetic, functional and mechanical properties, application and advantages/disadvantages of the following thermoplastics in the production of graphic products and commercial packaging:

(followed by the list of specific polymers)

Here, you need to be familiar with the specific properties of the polymers listed, where they are best used and why. The stem is further clarified by the use of polymers in graphic products (e.g. point-of-sale, product casings) and commercial packaging only. A question will not be asked, for example, on why PVC is best used for drainpipes as this is not a graphic product.

The following are the main terms used in Unit 2.

Key term in section stem	Meaning
Aesthetic properties	The visual qualities of materials.
Functional properties	The qualities a material must possess in order to be fit for purpose, e.g. the correct weight, grade, size.
Mechanical properties	The material's reaction to physical forces, e.g. strength, plasticity, ductility, hardness, brittleness, malleability.
Application	The quality of being usable for a particular purpose or in a special way; relevance.
Advantages/disadvantages	Qualities and features favourable to success or failure.
Processes	A description of the systematic series of actions needed to produce something.
Structural composition	How a material is made up.
Characteristics	Recognisable features that help to identify or differentiate one process from another.
Preparation	Action required before a process can begin.
Production/manufacture	The process of manufacture.
Concept	The general idea behind the use of quality assurance systems.
Principles	The distinct reasons for health and safety legislation.

The following are the main terms used in Unit 3.

Key term in section stem	Meaning
Application	The quality of being usable for a particular purpose or in a special way; relevance.
Advantages/disadvantages	Qualities and features favourable to success or failure.
Processes	A description of the systematic series of actions needed to produce something.
Characteristics	Recognisable features that help to identify or differentiate one process from another.
Production	The process of manufacture.
Principles	The distinct reasons for something.
Impact	Effects felt as a result of man's intervention/modern systems.
Sources	Raw materials for processing.
Debate	Discussion involving opposing viewpoints.
Responsibilities	The duty and obligations of developed countries.

Unit 3: Designing for the Future

For example:

Computer integrated manufacture (CIM)

Characteristics, processes, application, advantages/ disadvantages and the impact on employment of CIM systems to integrate the processing of production and business information with manufacturing operations, including:

(followed by a list of characteristics of CIM systems)

Here, you must study a wide range of aspects relating to the use of CIM systems. Firstly, you need to develop an in-depth knowledge and understanding of the features of CIM systems and how they are used to produce products. Then, you must explain the advantages and disadvantages of using CIM systems, in particular their effect on the modern workforce.

Portfolio of Creative Skills

Summary of expectations

1 What to expect

In this unit you have the opportunity to develop your creative, technical and practical skills through a series of product investigation, design and manufacturing activities. The unit is divided into three different sections: product investigation, product design and product manufacture. Each of these sections is separate and is not dependent on the other two, as opposed to one extended coursework project. All of the skills developed in this unit will be put to great use in the full design and make exercise in **Unit 4: Commercial Design at A2 level**.

This unit is set and marked by your teachers, then sent to Edexcel for moderation (sampling and checking of teachers' marks).

Product investigation

In this section you are free to choose any appropriate product(s) that interest you for your product investigation, so long as there is the opportunity to develop your skills in examining product performance, materials and components, product manufacture and quality issues. Alternatively, the choice of product(s) may be set by your teacher to ensure that a range of materials, techniques and processes are covered.

Product design

When working on the product design section of Unit 1 you are not limited by the manufacturing or materials constraints of your school or college workshop. There is no requirement for your designs to be carried forward into a manufactured product. Therefore, you can design as openly as you like, developing creative and adventurous

design, modelling and communication skills. Modelling during the development stage should be photographed to provide evidence. The individual design briefs or needs can either be set by yourself or given to you as a design exercise by your teacher.

Product manufacture

In the product manufacture section you have the opportunity to develop many practical skills through making more than one product using a range of different materials. During product manufacture you do not need to design the product, as the focus is on gaining and developing practical abilities, plus those of planning for production and prototype testing.

Manufacturing briefs should be set by your teacher to ensure that you can target specific making skills and processes with a view to developing a broad set of skills and experiences in a variety of materials. Photographic records of all stages of manufacture are essential in providing evidence of advanced skills, level of difficulty and complexity so that you can access the marks you deserve.

2 What is a Graphic Product?

Graphic Products has two clearly defined pathways: either 'conceptual design' or 'the built environment'.

(i) **Conceptual design** incorporates a wide range of 3D products with associated graphics, for example:
   - packaging design
   - product/industrial design
   - point-of-sale display
   - vehicle design.

(ii) **The built environment** focuses on the surroundings that provide the setting for human activity, for example:

- architecture
- interior design
- exhibition design
- theatre sets
- garden design.

In this unit you can explore both pathways in order to evidence the assessment requirements for the **three** distinct sections.

3 How will it be assessed?

Unit 1 is divided into three main sections with several sub-sections to focus your work. Each of these separate sub-sections contains assessment criteria that are allocated a certain amount of marks. A breakdown of each assessment criterion will be outlined in the **'To be successful you will'** sections of this textbook.

The maximum number of marks available for each section is 30, with an overall mark out of 90.

Sections	Sub-sections	Marks
Product investigation	A. Performance analysis	6
	B. Materials and components	9
	C. Manufacture	9
	D. Quality	6
Product design	E. Design and development	18
	F. Communicate	12
Product manufacture	G. Production plan	6
	H. Making	18
	I. Testing	6
	Total marks:	**90**

4 Building a portfolio

You will submit **one** portfolio that contains evidence for all **three** distinct sections. Your portfolio should contain a variety of different pieces of work that covers a wide range of skills and demonstrates an in-depth knowledge and understanding of Graphic Products. Your portfolio can contain several separate investigating, designing and making tasks, or a few combined design and make tasks.

Example: separate investigating, designing and making tasks:

Product investigation	Product design	Product manufacture
Product investigation 1: Analysis of different perfume bottles and packaging.	**Design task 1:** (2D design) Design of a magazine cover using ICT.	**Making task 1:** Making a Styrofoam™ model of a games controller from a working drawing.
Product investigation 2: Disassembly of an mp3 player.	**Design task 2:** (3D design) Design of a hand-held games console.	**Making task 2:** Making an MDF block model of a hand-held games console (from Design task 2).
	Design task 3: (2D and 3D design) Design of a new ice-lolly and associated packaging.	**Making task 3:** Making a foam-board architectural model of a building and its interior based on your school/college site.

Example: combined design and make tasks:

Product investigation	Product design	Product manufacture
Product investigation 1: Research and analysis of perfume bottles and their packaging.	**Design task 1:** (2D and 3D design) Design of a new perfume bottle and associated packaging.	**Making task 1:** Making a model of a new perfume bottle and its packaging.
Product investigation 2: Research and analysis of mp3/CD players.	**Design task 2:** (3D design) Design of a 'next generation' portable music system.	**Making task 2:** Making an MDF model of the 'next generation' portable sound system.

5 How much is it worth?

The portfolio of creative skills is worth 60 per cent of your AS qualification. If you go on to complete the whole course, then this unit accounts for 30 per cent of the overall full Advanced GCE.

Unit 1	Weighting
AS level	60%
Full GCE	30%

Product investigation (30 marks)

Getting started!

In this section, you will analyse a range of existing commercial products using your knowledge and understanding of designing and making. You should take into consideration the intended function and performance of the product; the materials, components where appropriate and processes used during its manufacture; how it was produced and how its quality was assured.

FACTFILE:

When setting product investigation tasks you must take into account the following.

- Your chosen product must contain **more** than one material and process in order to access the full range of marks.
- You may investigate a range of different products over the course of your AS studies. However, for your portfolio, evidence of only **one** complete product investigation should be submitted. Evidence must **not** comprise the best aspects of a range of product investigations that you have undertaken.
- The submitted product can be chosen by you or by your teacher.

A Performance analysis (6 marks)

When analysing your chosen product, you should determine what it was that the designer set out to achieve and then produce a technical specification that covers several key headings.

The technical specification should include the following.

- **Form** – why is the product shaped/styled as it is?
- **Function** – what is the purpose of the product?
- **User requirements** – what qualities make the product attractive to potential users?
- **Performance requirements** – what are the technical considerations that must be achieved within the product?
- **Material and component requirements** – how should materials and components perform within the product?
- **Scale of production and cost** – how does the design allow for scale of production and what are the considerations in determining cost?

Your specification points should contain more than a single piece of information, so that each statement is fully justified by giving a reason for the initial point. For example, it is not sufficient to say 'the material used is polystyrene', as this is not justified until 'because it is tough and can be injection moulded' is added.

As part of this analysis, you should also look at one other existing similar product, using the same criteria identified in your technical specification. By finding out information on a similar product you can compare and contrast it with your own chosen product.

ACTIVITY:

Start your product investigation by carrying out a detailed physical study of your product. This will enable you to look at the product in closer detail and provide an opportunity to develop your communication skills.

1. Sketch a 3rd angle orthographic view of the product and, using Vernier callipers and/or a micrometer, accurately record the most important dimensions.
2. Construct an accurate 3rd angle orthographic drawing (to a suitable scale) of the product using a technical drawing board and equipment. Use British Standard dimensioning and labelling.
3. Draw the product in three dimensions using a pictorial drawing method such as isometric or two-point perspective. Use studio markers or coloured pencils to render the drawing to provide a realistic representation of the product.

Note: you could use suitable computer-aided design (CAD) software to perform tasks 2 and 3 in order to develop your ICT skills.

Performance analysis: perfume bottles

 PRODUCT 1: DKNY Red Delicious	CRITERIA	 PRODUCT 2: Dali Kiss	ANALYSIS
The product is styled to resemble an apple – a play on the type of apple with the same name. The glass is coloured red to reflect the 'red delicious' brand. A 'red delicious' sticker is clearly visible to resemble the sticker on a real apple.	Form	The product is styled to resemble one of Salvador Dali's famous surrealist paintings and takes the form of a nose for the closure and lips for the bottle shape.	Both products use unusual forms to appeal to the consumer. Something out of the ordinary and not simply a normal bottle shape – individual and quirky.
The function of this product is to contain 100ml of fragrance securely by using a glass container, which is inert. The fragrance is dispensed by means of an atomiser that is activated by a push button on the top of the container.	Function	The function of this product is to contain 100ml of fragrance securely by using a glass container, which is inert. The fragrance is dispensed by means of an atomiser that is protected by a plastic closure for hygiene.	Both products have the same basic function but their form means that they operate in slightly different ways.
The size and shape makes it easy to hold in the hand and it is ergonomically sound so that it is easy to activate the push button to dispense the fragrance. The highly polished top surface complements the red glass and would look great on the customer's dressing table or bathroom shelf as a statement of their style.	User requirements	The width of the lips makes it slightly less comfortable to hold and dispense. The removable closure is another component that could be misplaced or damaged, which would detract from the overall style of the product.	Apart from the actual smell of the fragrance, the bottle shape is extremely important in attracting the attention of the consumer. Both would appeal to the consumer.
The atomiser must be activated when the button on the top of the container is pressed. Therefore, the internal mechanism must link this movement into a forward release of vapour. The highly polished component must fit perfectly on top of the glass container, which should be sealed to prevent liquid escaping.	Performance requirements	The glass container must be sealed with an appropriate atomiser system to enable the fragrance to be dispensed. The lid must fit securely over the atomiser to prevent it from coming loose.	Both products use fairly basic atomiser systems that enable the efficient and accurate dispensing of fragrance. Accurate production of components enables high-quality assemblies.
Glass is used because it is inert so it doesn't react with the fragrance. It is extremely tough and will not shatter easily, whilst glass gives a high-quality feel to the product, is crystal clear and can be coloured for effect. Polystyrene is used for the lid as it is tough and easily moulded. A highly polished silver effect is sprayed on to the grey plastic for visual appeal.	Materials & components requirements	Glass is used because it can be easily moulded into complex shapes and coloured and frosted for visual appeal. Frosted polystyrene is used for the lid as it can be easily moulded and will not shatter easily for this removable component.	Glass is the ideal material for containing perfumes. Visual appeal can be added by using plastic components. Both products use an atomiser system.
The glass container is produced using the automatic blow and blow process, which is ideal for large-batch and mass production, satisfying consumer demand. The lid is injection moulded for the same reasons as the container. The cost is £29.99 for 100ml, which is expensive due to the designer brand and not the actual cost of materials and production.	Scale of production & cost	The glass container and lid are manufactured in the same way as the DKNY product. The cost is £14.99 for 100ml as the brand is not as well developed.	Perfumes command a high price due to designer labels and consumer demand for luxury products.

Figure 1.1 An example of performance analysis for the comparison of two perfume bottles.

To be successful you will:

Assessment criteria: A. Performance analysis

Level of response	Mark range
Fully justify key technical specification points (1 mark) that relate to form, function, user requirements, performance requirements, materials and/or component requirements, scale of production and costs. (1 mark) Compare and contrast one other existing similar product using the technical specification. (1 mark)	4–6
Identify (1 mark) with some justification (1 mark) a range of realistic and relevant specification points that include reference to form, function and user requirements. (1 mark)	1–3

Marks are awarded in the following order: you must achieve all the marks from the lower section first (1–3) before being awarded marks from the higher section (4–6). This applies to all the assessment criteria.

B Materials and components (9 marks)

You will need to identify the materials and components used in your chosen product and apply your knowledge and understanding of their properties and qualities to suggest why in particular they have been selected for use. For example, cartonboards are used extensively in the retail packaging industry, where specific properties are required. These boards must be suitable for high-quality, high-speed printing and for cutting, creasing and gluing using very-high-speed automated packaging equipment.

Advantages of using cartonboard include:
- total graphic coverage and excellent print quality
- excellent protection in structural packaging nets
- relatively inexpensive to produce and process
- can be recycled.

LINKS TO:

Unit 2: Materials and components will provide you with the majority of information required to determine the choice of materials and components for a range of graphic products.

As with many products, your chosen product could have been made effectively in terms of quality and performance from other materials and components. Therefore, you should investigate suitable alternative materials and components and, using advantages and disadvantages, compare them with the materials and components actually used. For example, packaging can be made from a wide variety of materials, not just cartonboard as stated earlier. Companies may use metals, polymers or even woods to package their products, especially expensive gifts where they are trying to portray a sophisticated image.

ACTIVITY:

In groups of three, each select one different type of packaging used for the same kind of product. For example, fragile products such as biscuits can be packaged in card cartons, metal containers or plastic tubs – all with added internal packaging. Each person has to carry out an analysis of the materials and components used for their selected type of package.

Still in groups of three, each person has to explain why their type of package is better for containing the product. This mini debate should uncover the advantages and disadvantages of the materials used and you should be able to arrive at a joint decision as to the most suitable material overall.

Sustainability is an important aspect to consider with any product and you should be able to explain the environmental effects of using the materials identified in the product in relation to one or more of the following.

- **Extraction and processing of raw materials**– what is the financial and environmental cost of using a particular material in terms of energy use and pollution? Can less material be used? Can recycled materials be used or can the product be designed so that it is easily recyclable?
- **Production processes** – do they require lots of energy or produce lots of waste products? Can the product be simplified to reduce the amount of production processes?
- **Disposal of products after their useful lifespan** – does the product minimise waste production? (Reduce, re-use, recover and recycle.)

LINKS TO:

Unit 3: Sustainability looks at these issues in greater detail. Although it is an A2 unit, the information given will be extremely helpful to this section.

Materials and components analysis: DKNY 'Red Delicious' perfume bottle

Component	Why selected	Possible alternatives	Sustainability issues
Glass container	Glass is an inert material that will not react with the fragrance contained inside. Glass can be formed into complex and interesting shapes such as the apple form by the automatic blow and blow process, which also makes it suitable for large-batch or mass production. Glass has a high-quality appearance that is crystal clear and can be coloured red to fit with the 'red delicious' theme. Glass is impervious to water and air, which enables the container to be sealed for use with an atomiser dispensing system.	It is hard to suggest any real alternatives to glass for containing a fragrance due to its excellent functional, mechanical and aesthetic properties. Polymers can be used to provide decoration around a glass container as in the design of some perfume bottles.	Large amounts of energy are required to process glass from its raw materials through the automatic blow and blow process due to high temperatures needed to melt glass. Glass can be recovered and recycled after use. Alternatively, some recycled glass content could be used without degradation of overall quality. The glass container cannot be refilled due to the atomiser system.
Polystyrene lid	Polystyrene is a thermoplastic and can therefore be easily shaped using thermoforming techniques (injection moulding). Polystyrene has very good functional and mechanical properties including strength, toughness, durability and resistance to impact and water, which make it suitable for this application. The grey polystyrene (with some recycled content) can take a high-quality surface finish such as the chrome effect created by dip coating. Polystyrene is lightweight, so it does not add any significant weight to this glass container.	There are a couple of alternatives to polystyrene for this application including ABS and polycarbonate. However, as polystyrene possesses the majority of properties required and is the less expensive alternative, it is ideal for the lid of the perfume bottle.	As with most polymers, polystyrene can be recycled if the components can be separated easily for sorting. The separation of the lid from the glass container would be problematic, so a better design should be developed to enable this. The chrome effect would also prove problematic to remove even when separated. A silver polystyrene could be used but it wouldn't look as good. Polymers are derived from a finite resource – oil – and are not sustainable. They also take hundreds of years to degrade in landfill. A biodegradable polymer could be used but they are, at present, not sufficiently developed for this application.
Printed vinyl sticker	PVC film provides a thin yet durable substrate for adding surface graphics. PVC has excellent print quality to advertise the DKNY brand in full colour. Shiny surface finish that creates a high-gloss effect, which fits with other reflective surfaces of chrome and glass.	A printed paper label could be used in place of the PVC one. However, the paper may become wet and peel off after prolonged use. PVC is water resistant.	PVC is quite polluting to the environment due to hydrochloric acid being produced when incinerated. However, PVC takes less energy to produce than other polymers. The printing inks used would most likely be solvent based, which cause volatile hydrocarbons to be released into the atmosphere. Water-based inks should be used instead.
Atomiser system	An atomiser system is utilised to dispense a measured amount of fragrance after each press of the button on the lid. The system provides a seal to the neck of the glass container that prevents any liquid from escaping and external contaminants from entering.	The perfume could be eau de toilette, which means it would not need an atomiser system. However, the design of the bottle would have to be developed so that the lid screwed off the glass container to gain access to the open neck.	An atomiser system does not use compressed gases as in the case of an aerosol, so no dangerous chlorofluorocarbons (CFCs) are released into the atmosphere. CFCs have been proven to contribute to the depletion of the ozone layer, which contributes to global warming.

Figure 1.2. *An example of the analysis of the materials and components used in a perfume bottle.*

To be successful you will:

Assessment criteria: B. Materials and components

Level of response	Mark range
Suggest, with reference to quality and performance, alternative materials and/or components that could have been used in the product. **(1 mark)** Evaluate, using advantages and disadvantages, the selection of the materials and/or components used. **(1 mark)** Describe the impact on the environment of using the materials and/or components identified. **(1 mark)**	7–9
Describe a range of useful properties that relate to the materials and/or components identified **(1 mark)** and justify their selection and use in the product. **(1 mark)** Identify alternative materials and/or components that could have been used in the product. **(1 mark)**	4–6
Identify a material or component used in the product. **(1 mark)** Describe a useful property of that material or component **(1 mark)** and justify its use. **(1 mark)**	1–3

C Manufacture (9 marks)

You need to be able to identify, describe and justify the processes involved in the manufacture of your chosen product. For example, the polymer casings for many electrical products are made using the injection moulding process. This is due to many reasons, including its suitability for mass production, its speed and accuracy and the ability to create complex and interesting shapes the consumer may find appealing.

It is important to consider that other methods of manufacture could have been used, so you should make clear **one** alternative method and compare and contrast it with the actual methods used. For example, the polymer casings for many mp3 players result in 'cheap'-looking products. Some companies, however, use anodised aluminium or polished stainless steel casings that add 'weight' and have a higher quality 'feel' to them. Obviously, injection moulding cannot be used for metals so alternative processes such as extrusion and press forming need to be employed.

LINKS TO:

Unit 2: Industrial and commercial practice will provide you with many manufacturing processes used to produce a range of graphic products.

ACTIVITY:

Study an example of the printed materials produced commercially for your school or college, i.e. brochure, prospectus, yearbook, etc. Determine what commercial printing processes, printing effects and binding methods have been used to produce these materials.

Now, think about alternative methods that could be used to produce those printed materials 'in-house' by you and your classmates. How could you produce identical copies using the printing and binding resources available to you, such as digital printing or colour photocopying, laminating or photo-glossy paper, thermal binding or spiral binding?

Again, you should also consider and describe the effects that using particular commercial processes in your product has on the environment. For example, the aim of many manufacturers is to reduce production costs by creating designs that use less material and less energy during manufacture and to reduce waste production. Controlling emissions of harmful substances during production is also a serious consideration, e.g. carbon dioxide produced from the burning of fossil fuels for energy.

LINKS TO:

Unit 3: Sustainability looks at these issues in greater detail. Although it is an A2 unit, the information given will be extremely helpful to this section.

Manufacture analysis: Frijj milkshake bottle

Component 1: Closure (screw top lid)
Material: LDPE
Manufacture: Injection moulding

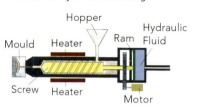

Justification:
- ideal for mass production – low unit cost for each moulding for high volumes
- precision moulding for screw thread and ribs on cap used for grip.

Possible alternative:
- blow moulding or vacuum forming could be used if the closure was a snap-on kind (did not require internal screw-thread).

Sustainability issues:
- process requires heating polymers so lots of energy use, although very little waste is produced.

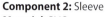

Component 2: Sleeve
Material: PVC
Manufacture: Shrink wrapping

Justification:
- high-quality decoration allows for full-body 360° labelling for high impact on shelf
- suited for tamper-evident seals (freshness seals/security seals)
- provides decoration on unusual shaped bottles using flexography and gravure printing.

Possible alternative:
- polypropylene shrink wrap
- printing directly onto bottle using rotary screen-printing process.

Sustainability issues:
- shrink wrapping requires PVC to be heated to 58 degrees, which requires a lot of energy in continuous production.

Component 3: Bottle
Material: HDPE
Manufacture: Blow moulding

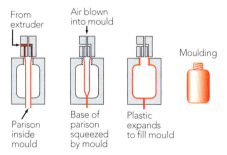

Justification:
- contoured shape of bottle can easily be formed
- hollow bottle shape with thin walls to reduce weight and material costs
- ideal for mass production – low unit cost for each moulding.

Possible alternative:
- none – blow moulding is the most efficient process for mass-producing a polymer bottle.

Sustainability issues:
- process requires heating polymers so lots of energy use although very little waste is produced.

Component 4: Printed graphics
Manufacture: Flexography

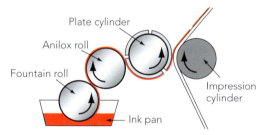

Justification:
- high-speed printing process suitable for continuous production
- fast-drying ink means high-quality printing produced
- relatively inexpensive to set up, therefore increased profits.

Possible alternative:
- Gravure – especially suitable if a perforated seal was required around the neck of the bottle instead of the snap-off polymer seal used
- digital UV flatbed printing only suitable for smaller runs.

Sustainability issues:
- solvent-based printing inks are used, which are not as 'green' as vegetable-based printing inks.

Figure 1.3 *An example of the analysis of the manufacturing processes used for a drinks bottle.*

To be successful you will:

Assessment criteria: C. Manufacture

Level of response	Mark range
Evaluate, using advantages and disadvantages, the selection of the manufacturing processes used in the product. **(1 mark)** Suggest one alternative method of production that could have been used in the manufacture of the product. **(1 mark)** Describe the impact on the environment of using the processes identified in the production of the product. **(1 mark)**	7–9
Describe: **(1 mark)** • a range of processes used in the manufacture of the product **(1 mark)** • and fully justify their use for the level of production of the product. **(1 mark)**	4–6
Identify, **(1 mark)** describe **(1 mark)** and justify the use of a manufacturing process used in the construction of the product. **(1 mark)**	1–3

D Quality (6 marks)

All products will have gone through a series of checks and tests to ensure they reach the consumer in the best possible condition in terms of quality and performance. You need to describe when and where quality control (QC) checks take place during the manufacture of your product, what the checks consist of and how they form part of a quality assurance (QA) system. For example, all printed materials will have undergone QC using printers' marks, including colour bars and registration. Many of these marks may still be visible on the products you investigate, e.g. some frozen food or cereal cartons contain colour bars and registration marks on their tabs. However, printers' marks are usually printed outside the margins of printed products and are subsequently cut off.

You also need to identify and describe some of the main external standards that must be met during product manufacture and how they influence production and the final product. For example, what British Standards need to be adhered to when producing your product? These can cover a wide range of topics including materials selection, individual component testing/overall product testing or standard of service and management. Some of these may result in the awarding of the British Standards Institute (BSI) 'Kitemark' or the European CE mark for quality assurance.

FACTFILE:

Quality assurance (QA) is the system used by the manufacturer to monitor the quality of a product from its design and development stage, through its manufacture, to its end-use and the degree of customer satisfaction. In other words, QA is an assurance that the end product fulfils all of its requirements for quality.

Quality control (QC) is part of the achievement of QA. It involves the actual inspection and testing activities used by a manufacturer to ensure a high-quality product is produced.

External quality standards are used when testing, inspecting and verifying the overall quality of materials, components, products and systems. These **formal standards** are produced through standards organisations for national (BS), European (EN) or international (ISO) use.

LINKS TO:

Unit 2: Quality will provide you with the necessary information on issues relating to QC and QA procedures used in the production of Graphic Products.

To be successful you will:

Assessment criteria: D. Quality

Level of response	Mark range
Describe a range of QC checks used during the manufacture of the product **(1 mark)** and explain how the main relevant standards influenced the manufacture of the product. **(1 mark)** Describe a QA system for the product. **(1 mark)**	4–6
Identify, **(1 mark)** describe **(1 mark)** and justify the use of one QC check during the manufacture of the product. **(1 mark)**	1–3

Quality control for electronic components.

- Electronic components would be bought-in from specialist electronics companies. Microsoft® engineers would monitor levels of quality at source and agree manufacturing specifications and tolerances suitable for this product.
- Sampling and testing of electronic components using bench tests to determine performance.
- Sampling and testing of circuit board assemblies looking for dry joints, which would prevent electrical current from flowing through circuit.
- Soak testing of final assemblies to determine performance, over-heating problems, etc.

Electronic components (QC) **1**

Quality control for injection-moulded casing.

- Polystyrene pellets coming into factory would be checked against manufacturing specifications so that the correct grade and colour of material is being used.
- The injection mould would be visually inspected periodically for any damage or dirt/grit that would affect quality of casing. Computer analysis would be far more efficient at determining very small defects but mould would have to be taken out of machine (substituted).
- Sampling and testing of batches of injection-moulded components against agreed tolerances. Fine tuning of machinery if sampling chart shows tolerance creep.
- Sampling and testing of assembly of two halves of split casing to ensure that they accurately snap together.

Injection moulding casing (QC) **2**

Quality assurance for Xbox 360 games controller.

Quality assurance (QA) **3**

Preparation
- Raw materials such as polystyrene for casing, solder, etc., would all be sourced from reputable suppliers and quality checks made on a regular basis on batches of orders.
- Electronic components would be outsourced to specialist electronics companies who would have to meet Microsoft's tolerances for component manufacture.

Processing
- QC checks feature heavily in manufacture of electronic components and injection moulding of casing.
- QC checks on screen-printing of surface graphics.
- Sampling and testing of batches of components and casings against agreed tolerances.

Assembly
- Sampling and testing of sub-assemblies of circuit boards, split casings and button assemblies against agreed tolerances.
- Soak testing of batches of final assemblies to determine final quality of product – is it fit-for-purpose?

Finishing
- Packaging product/collating into corrugated board boxes/palletising – checking quantities for dispatch.

After-sales
- Guarantee assures customer that the games controller will be fit-for-purpose and will not break under normal conditions.

Figure 1.4 *An example of the analysis of quality issues related to a computer games controller.*

Product design (30 marks)

Getting started!

In this section, you will have the opportunity to demonstrate your creativity and flair by using your design skills, through the production of a range of alternative ideas that explore different approaches to a problem. Using the best aspects of your initial designs, you will develop and refine your ideas, with the aid of modelling, into a final workable design proposal that will satisfy a design brief or specific need.

Your designs do not have to be manufactured but the most viable products must be communicated to potential users. Any designer must sell their ideas by the use of presentation graphics or concept boards. Accurate working drawings and assembly drawings provide an audience with technical details of the product. Both forms of communication are invaluable in presenting an impression of the final product.

FACTFILE:

When setting product design tasks you must take into account the following:
- you can respond creatively and adventurously to one or more design briefs or needs
- design briefs or needs can be set either by you or by your teacher to produce solutions that are both fit for purpose and market viable
- design tasks can be explored in 3D products/environments and 2D 'graphic design' of printed materials but the portfolio must have evidence of both.

E Design and development (18 marks)

There are a number of possible starting points to this section. The design brief or need may be given to you by your teacher or you may define your own.
Two possible types of brief that you might want to use are:
- a focused design brief for a specific need or want
- a 'blue sky' project resulting in concepts using future technologies.

Design brief

Design a leaflet that challenges the stereotypical views of how the young perceive the older generation and how the old perceive the younger generation. The leaflet is to be delivered with a local newspaper and be backed by the local police force in order to promote 'safer neighbourhoods'.

The leaflet should be made from a single sheet of A4 paper, with two folds resulting in three columns to a side, i.e. tri-fold. The information communicated should challenge stereotypes in a humorous manner but not cause offence to either age group. The leaflet should be designed using desktop publishing (DTP) software and be full-colour printed.

Design brief

Design an interactive information kiosk for an airport departure lounge including interactive information screens.

Design specification
- The user should be able to access travel information, book hotels/hire cars with a credit card and enter secure areas of personal/business information using biometric security systems.
- The kiosk must look contemporary in order to fit into a modern airport departure lounge.
- The kiosk should be accessible to able and disabled users in wheelchairs.
- The kiosk should occupy a floor space of no more than 1m × 1m.

Outcomes
- A series of design sheets developing the 3D kiosk and 2D information screens.
- A 3D Styrofoam™ model of the kiosk made to an appropriate scale.
- A sample of three interactive information screens, including the main 'welcome' screen, using ICT.

Figure 1.5 Examples of possible types of briefs.

You are not required to write a detailed design specification for each design task. However, the design brief must contain a range of design criteria that your final design proposal must meet. Therefore, you need to consider the design problem set and produce a range of alternative design ideas that focus on the whole or parts of the problem. It is not necessary for you to produce a wide range of alternative ideas. It is better to produce high-quality focused work than lots of lower-quality work.

Throughout your work you should explore different design approaches, applying your knowledge of materials, components, processes and techniques to produce realistic design proposals that satisfy the design brief or need. Design ideas should be objectively evaluated against the criteria set out in the design brief or need, to ensure that

your designs are realistic and viable. The use of detailed annotation is an important feature of design development and you should use it to explain details of design thinking and to offer thoughts on your design proposals.

It is important that you develop your own individual style when designing. Not everyone can produce beautifully presented and professional looking design sheets – the important thing is that you effectively communicate your design intentions. Experiment with a range of studio materials such as sketching with pencils, fine-liners, etc., on different types of papers, e.g. ballpoint pen on tracing paper or white pastel pencil on brown paper. Look at how professional designers present their design ideas and try to develop a more 'designerly' approach than your GCSE coursework projects.

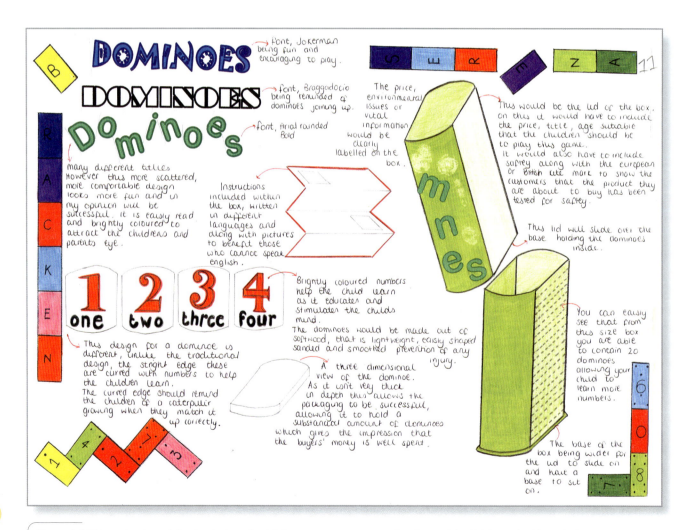

Figure 1.6 *This student's work for the design of a child's game is very structured and resembles GCSE work, although designs are communicated very well.*

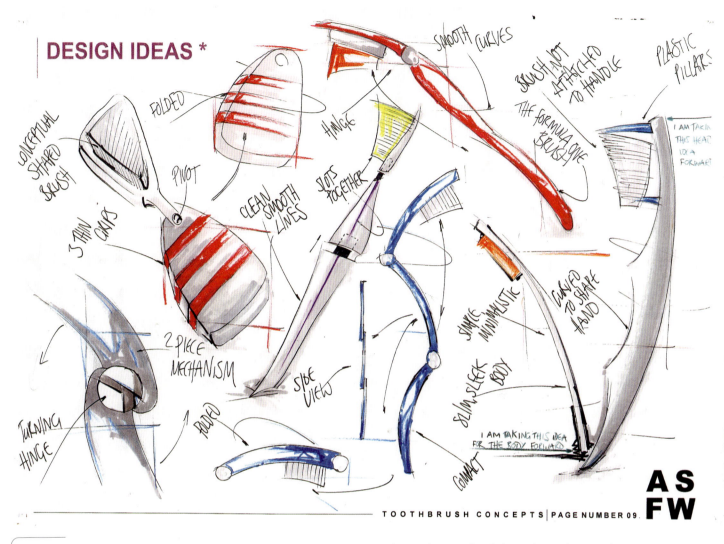

DESIGN IDEAS *

CONCEPTUAL SHAPED BRUSH

FOLDED

SMOOTH CURVES

PIVOT

HINGE

BRUSH NOT ATTACHED TO HANDLE

PLASTIC PILLARS

THE FORMULA ONE BRUSH

3 THIN CHIPS

CLEAN SMOOTH LINES

SLOTS TOGETHER

I AM TAKING THIS HEAD IDEA FORWARD

2 PIECE MECHANISM

SIMPLE MINIMALISTIC

CURVED TO SHAPE HAND

TURNING HINGE

FOLDED

SIDE VIEW

SLIM SLEEK BODY

I AM TAKING THIS IDEA FOR THE BODY FORWARD

COMPACT

A S FW

TOOTHBRUSH CONCEPTS | PAGE NUMBER 09

Figure 1.7a *This student's work for the design of a toothbrush adopts a more professional approach with 'busy' design sheets and highlighting using spirit-based markers.*

ACTIVITY:

To practise your designing skills, set yourself a few small and manageable design tasks. These tasks should be focused and limited to three hours of design time. For example:

- Design an in-car satellite navigation system that can be detached and used when walking. The product should be easily installed in a car and removed for use when walking, be easy to read and ergonomically sound to handle.

- Design the front cover of a new paperback book entitled *Terror at 30,000 Feet* using ICT. The cover must measure 17cm (width) × 23cm (height), include the title of the book, its author, J.R. Hartley, and present suitable imagery using a photo-composition to entice readers.

DEVELOPMENT *

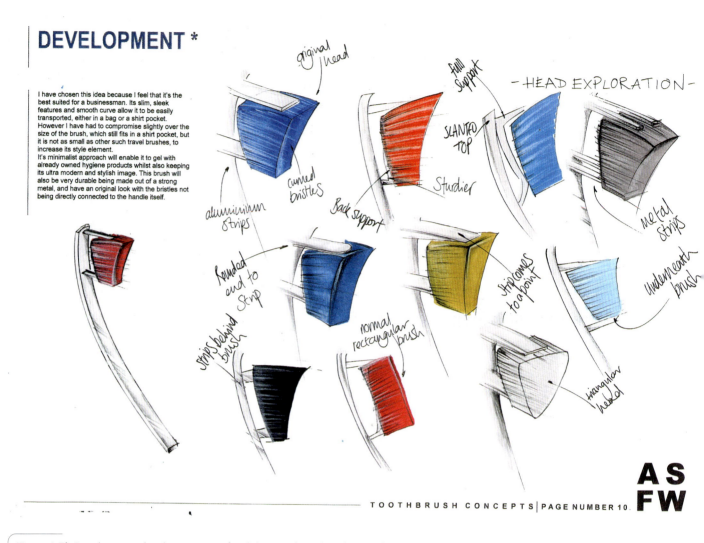

I have chosen this idea because I feel that it's the best suited for a businessman. Its slim, sleek features and smooth curve allow it to be easily transported, either in a bag or a shirt pocket. However I have had to compromise slightly over the size of the brush, which still fits in a shirt pocket, but it is not as small as other such travel brushes, to increase its style element.

It's minimalist approach will enable it to gel with already owned hygiene products whilst also keeping its ultra modern and stylish image. This brush will also be very durable being made out of a strong metal, and have an original look with the bristles not being directly connected to the handle itself.

TOOTHBRUSH CONCEPTS | PAGE NUMBER 10.

AS FW

Figure 1.7b *Development sketches contain refined drawings based on the initial design ideas.*

When developing your initial design ideas, the following design development cycle can be used. Development is an important part of the design process and should be used to refine an initial idea into a workable design solution.

Modelling should be used to test features such as proportions, scale, function, sub-systems, etc. Modelling can be achieved through the use of traditional materials or 2D and 3D computer simulations. Evidence of 3D modelling should be presented using clear, well-annotated photographs. Card mock-ups or rough layouts of 2D printed materials should be included in your portfolio with associated evaluative comments clearly labelled.

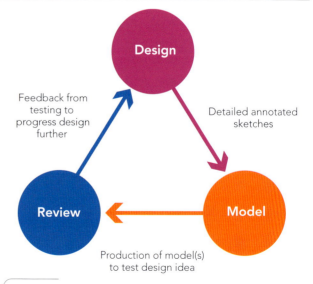

Figure 1.8 *Design development cycle.*

development of lounge

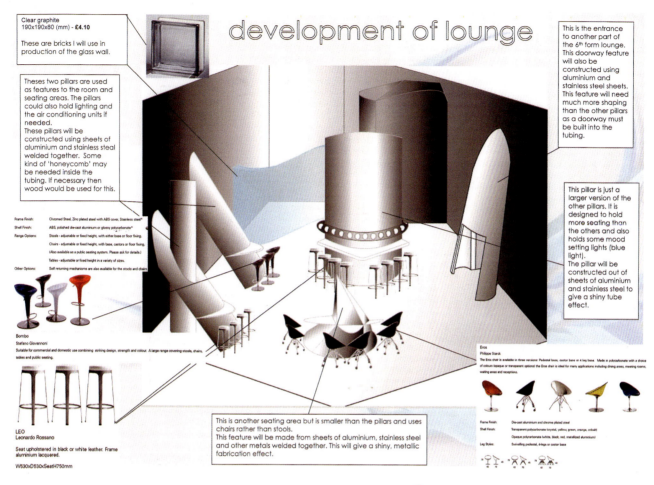

Clear graphite
190x190x80 (mm) - £4.10

These are bricks I will use in production of the glass wall.

Theses two pillars are used as features to the room and seating areas. The pillars could also hold lighting and the air conditioning units if needed.
These pillars will be constructed using sheets of aluminium and stainless steal welded together. Some kind of 'honeycomb' may be needed inside the tubing. If necessary then wood would be used for this.

This is the entrance to another part of the 6th form lounge. This doorway feature will also be constructed using aluminium and stainless steel sheets. This feature will need much more shaping than the other pillars as a doorway must be built into the tubing.

This pillar is just a larger version of the other pillars. It is designed to hold more seating than the others and also holds some mood setting lights (blue light).
The pillar will be constructed out of sheets of aluminium and stainless steel to give a shiny tube effect.

Frame Finish: Chromed Steel, Zinc plated steel with ABS cover, Stainless steel*
Shell Finish: ABS, polished die-cast aluminium or glossy polycarbonate*
Range Options: Stools - adjustable or fixed height, with either base or floor fixing.
Chairs - adjustable or fixed height, with base, castors or floor fixing.
(Also available as a public seating system. Please ask for details.)
Tables - adjustable or fixed height in a variety of sizes.
Other Options: Self returning mechanisms are also available for the stools and chairs.

Bombo
Stefano Giovannoni
Suitable for commercial and domestic use combining striking design, strength and colour. A large range covering stools, chairs, tables and public seating.

LEO
Leonardo Rossano

Seat upholstered in black or white leather. Frame aluminium lacquered.

W530xD530xSeatH750mm

This is another seating area but is smaller than the pillars and uses chairs rather than stools.
This feature will be made from sheets of aluminium, stainless steel and other metals welded together. This will give a shiny, metallic fabrication effect.

Eros
Philippe Starck
The Eros chair is available in three versions: Pedestal base, castor base or 4 leg base. Made in polycarbonate with a choice of colours (opaque or transparent options) the Eros chair is ideal for many applications including dining areas, meeting rooms, waiting areas and receptions.

Frame Finish: Die-cast aluminium and chrome plated steel
Shell Finish: Transparent polycarbonate (crystal, yellow, green, orange, cobalt)
Opaque polycarbonate (white, black, red, metallized aluminium)
Leg Styles: Swivelling pedestal, 4-legs or castor base

different colour schemes

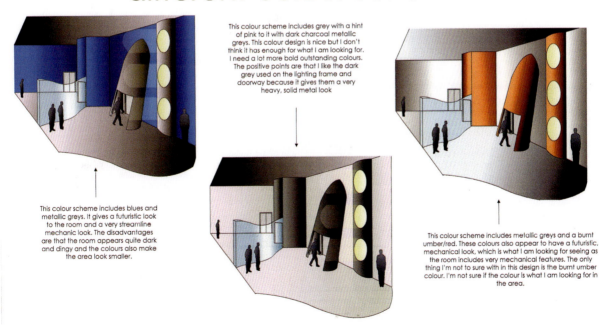

This colour scheme includes grey with a hint of pink to it with dark charcoal metallic greys. This colour design is nice but I don't think it has enough for what I am looking for. I need a lot more bold outstanding colours. The positive points are that I like the dark grey used on the lighting frame and doorway because it gives them a very heavy, solid metal look

This colour scheme includes blues and metallic greys. It gives a futuristic look to the room and a very streamline mechanic look. The disadvantages are that the room appears quite dark and dingy and the colours also make the area look smaller.

This colour scheme includes metallic greys and a burnt umber/red. These colours also appear to have a futuristic, mechanical look, which is what I am looking for seeing as the room includes very mechanical features. The only thing I'm not to sure with in this design is the burnt umber colour. I'm not sure if the colour is what I am looking for in the area.

Figure 1.9 *A student's development of an interactive information kiosk involving Styrofoam™ modelling using an ergonome to test both form and function.*

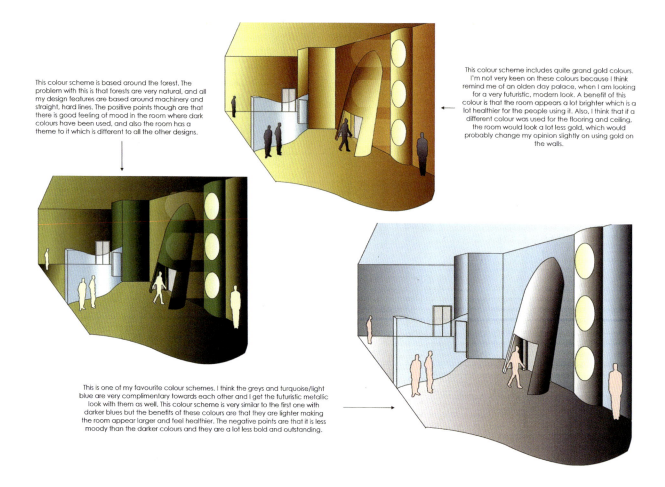

This colour scheme is based around the forest. The problem with this is that forests are very natural, and all my design features are based around machinery and straight, hard lines. The positive points though are that there is good feeling of mood in the room where dark colours have been used, and also the room has a theme to it which is different to all the other designs.

This colour scheme includes quite grand gold colours. I'm not very keen on these colours because I think remind me of an olden day palace, when I am looking for a very futuristic, modern look. A benefit of this colour is that the room appears a lot brighter which is a lot healthier for the people using it. Also, I think that if a different colour was used for the flooring and ceiling, the room would look a lot less gold, which would probably change my opinion slightly on using gold on the walls.

This is one of my favourite colour schemes. I think the greys and turquoise/light blue are very complimentary towards each other and I get the futuristic metallic look with them as well. This colour scheme is very similar to the first one with darker blues but the benefits of these colours are that they are lighter making the room appear larger and feel healthier. The negative points are that it is less moody than the darker colours and they are a lot less bold and outstanding.

To be successful you will:

Assessment criteria: E. Design and development

Level of response	Mark range
Present alternative ideas that are workable, realistic and detailed and that fully address the design criteria. **(1 mark)** Demonstrate detailed understanding of materials, processes and techniques in your ideas. **(1 mark)** Produce a final design proposal that is significantly different and improved compared with any previous alternative design ideas. **(1 mark)** Include technical details of materials and components, processes and techniques in the design proposal. **(1 mark)** Use modelling with traditional materials or 2D and/or 3D computer simulations to test important aspects of the final design proposal. **(1 mark)** Evaluate the final design proposal objectively against the design criteria in order to fully justify the design decisions taken. **(1 mark)**	13–18
Present realistic alternative design ideas. **(1 mark)** Present ideas that are detailed and address most design criteria. **(1 mark)** Use appropriate developments and use details from ideas to change, refine and improve the final design proposal. **(1 mark)** Present a final detailed design proposal. **(1 mark)** Use modelling to test some aspects of the final proposal against relevant design criteria. **(1 mark)** Objectively consider some aspects of the design brief/need in evaluative comments. **(1 mark)**	7–12
Present simplistic alternative design ideas. **(1 mark)** Present superficial ideas that address limited design criteria. **(1 mark)** Show developments that are minor and cosmetic. **(1 mark)** Present a basic final design proposal. **(1 mark)** Use basic modelling to test an aspect of the design proposal. **(1 mark)** Make evaluative comments that are subjective and superficial. **(1 mark)**	1–6

F Communicate (12 marks)

When presenting your design and development work, it is essential that you communicate your ideas effectively. Evidence for this section can be found throughout the following areas:

(i) Through your design and development work.

You should show evidence of 'design thinking' using any form of effective communication that you feel is appropriate. However, you should try to use a range of skills that may include freehand sketching in 2D and 3D, cut and paste techniques and the use of ICT. It is important to demonstrate a high degree of graphical skill, which will be shown through the accuracy and precision of your work.

The development of ICT skills is essential to Graphic Products. You must be able to produce professional-looking printed materials such as items of packaging that closely replicate those available commercially. Therefore, graphic design using DTP, drawing/painting and image manipulation software should be explored. When using CAD, you should ensure that it is used appropriately, rather than for show. For example, specialist CAD software to produce 3D rendered images is likely to be more appropriately used as part of development or final presentation, rather than for initial ideas.

Figure 1.10 A student's development of an ice-lolly wrapper using ICT.

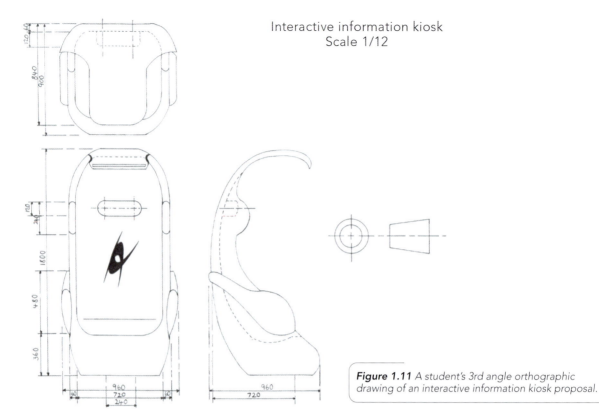

Interactive information kiosk
Scale 1/12

Figure 1.11 *A student's 3rd angle orthographic drawing of an interactive information kiosk proposal.*

(ii) Through your presentation graphics and technical drawings.

To effectively communicate final designs, a range of skills and drawing techniques should be demonstrated, which could include:

- **pictorial drawings** – isometric, planometric (axonometric), oblique and perspective drawings to convey a 3D representation of the product
- **working drawings** – 1st or 3rd angle orthographic, exploded assembly and sectional drawings to convey technical information
- **computer generated** – pictorial and working drawings, renderings, etc. using specialist software.

(iii) Through the quality of written communication.

Annotation should be used to explain design details and convey technical information. You should make sure that all information is presented in a logical order that is easily understood. Specialist technical vocabulary should be used consistently with precision.

To be successful you will:

Assessment criteria: F. Communicate

Level of response	Mark range
Use a range of communication techniques and media including ICT and CAD **(1 mark)**, with precision and accuracy **(1 mark)**, to convey enough detailed and comprehensive information to enable third-party manufacture of the final design proposal. **(1 mark)** Use annotation that provides explanation and most technical details of materials and processes with justification. **(1 mark)**	9–12
Use a range of communication techniques, including ICT **(1 mark)**, that are carried out with sufficient skill **(1 mark)** to convey an understanding of design and develop intentions and construction details of the final design proposal. **(1 mark)** Use annotation that provides explanation and most technical details of materials and process selection. **(1 mark)**	5–8
Use a limited range of communication techniques **(1 mark)** carried out with enough skill **(1 mark)** to convey some understanding of design and develop intentions. **(1 mark)** Use annotation that provides limited technical details of materials and processes. **(1 mark)**	1–4

Product manufacture (30 marks)

Getting started!

In this section, you will use your production planning skills and have the opportunity to develop your making skills through manufacturing **one or more** high-quality products to satisfy given design briefs or needs. You should use **a range** of materials, techniques and processes when manufacturing **a range** of products in order to build and develop a variety of skills and lay a foundation for more complex and challenging work in the future.

FACTFILE:

When setting product manufacture tasks you must take into account the following.

- You should produce **one or more** high-quality products that meet the requirements of the design briefs or needs.
- The design briefs or needs should contain requirements against which the final manufactured products can be measured.
- Some design briefs or needs may be set by your teacher to ensure a range of materials, techniques and processes are used.

There are a number of potential starting points to product manufacture.

- Making a product previously designed in the design and development section. This takes the combined design and make task approach to your portfolio.
- Making a product from a detailed working drawing and manufacturing specification provided by your teacher. Here your teacher will specifically target skills and materials that you need to evidence a wide range of techniques in your portfolio.
- The accurate replication or detailed modelling of an existing product or environment to a chosen scale. This could be linked to your product investigation task by accurately modelling the product you are studying.

ACTIVITY:

As an extension to the physical study task in the product investigation section, make a scale model of your chosen product from the information on your 3rd angle orthographic drawing.

Your model could be quickly made using Styrofoam™ as it is easy to cut and shape. However, for a more realistic model you could construct a block model using laminated medium-density fibreboard (MDF). This would enable you to apply a high-quality surface finish that might replicate the real thing.

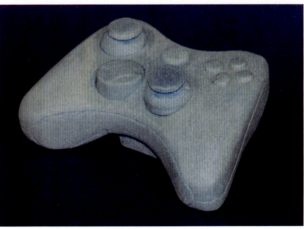

Figure 1.12 A student's Styrofoam™ model of the games controller used in the product investigation task.

Figure 1.13 *A student's interior model of a music rehearsal space for a school.*

LINKS TO:

Unit 3: Systems and control describes how flow charts are used to represent production processes.

G Production plan (6 marks)

You need to produce a detailed production plan that explains the sequence of operations carried out during the manufacture of each product. A production plan should contain a work order or schedule, which could be presented in the form of a flow chart. The work order should include the order of assembly of parts or components and tools, equipment and processes to be used during manufacture.

QC points should also be identified throughout the production plan in order for you to produce a high-quality product. Specific quality checks should be described and not simply stated as 'quality control'. Information regarding important safety checks may also form part of detailed planning.

An important part of planning is the efficient use of time, so you should make sure that you consider realistic timings and deadlines. Where Gantt or time charts are used, you must make sure that they are detailed, cover all aspects of product manufacture and include achievable deadlines.

Consideration should be given to the scale of production of your products. Although you may be making one-off products, most products would be batch or mass produced, so you should consider the consequences of these scales in your planning, developing your awareness of commercial production.

To be successful you will:

Assessment criteria: G. Production plan

Level of response	Mark range
Produce a detailed production plan **(1 mark)** that considers stages of production in the correct sequence **(1 mark)**, with realistic time scales and deadlines for the scale of production. **(1 mark)**	4–6
Produce a limited production plan **(1 mark)** that considers the main stages of manufacture **(1 mark)**, with reference to time and scale of production. **(1 mark)**	1–3

H Making (18 marks)

You should produce **one or more** high-quality products that meet the requirements of the design briefs or needs you have been given or developed yourself. The design brief or need must contain requirements against which the final manufactured product can be measured, so it is important when setting design requirements that they can be tested. Requirements may include dimensional parameters, finishes, etc., which are all objectively measurable requirements that can be tested for success.

TASK		5 hours per week			
		Week 1	Week 2	Week 3	Week 4
1	Laminating MDF	Drying times			
2	Cutting out rough profiles	½ hour			
3	Rough shaping of block	4½ hours			
4	Fine shaping of curves		4 hours		
5	Sanding smooth ready for finishing		1 hour		
6	Sealing MDF			Drying times	
7	Priming model			Drying times	
8	Finishing by applying top coat			Drying times	
9	Making wheels				1 hour
10	Assembling wheels				1 hour
11	Assembling whole model				1 hour
12	Applying finishing details				2 hours

TOTAL:	20 hours over 4 weeks

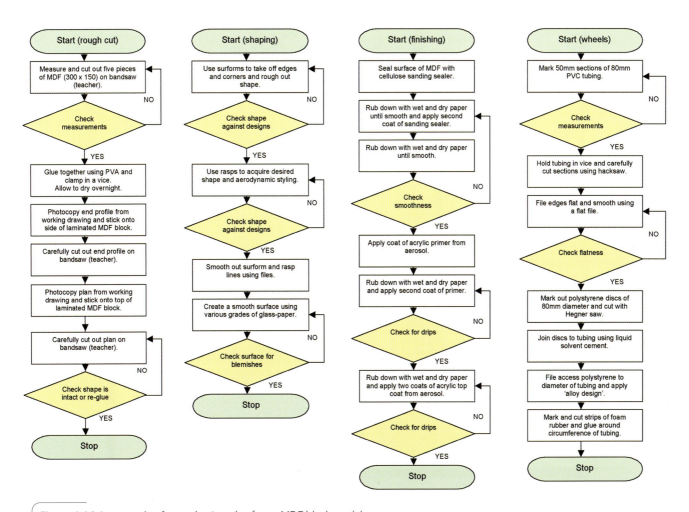

Figure 1.14 *An example of a production plan for an MDF block model.*

Design brief

You are required to make an accurate scale model of the Post-16 café that can be used to promote the school's Post-16 at careers fairs.

You are required to carry out a detailed site survey of the Post-16 café to determine dimensions and constructional details. The architectural plans for this building will be available to you to confirm measurements.

The model must be an accurate reproduction of the actual Post-16 café, including:

• a scale of 1:100
• a fully detailed exterior and interior
• a removable roof to view the interior layout.

Figure 1.15 An example of an appropriate design brief for manufacturing a model.

Throughout your making activities, you should demonstrate your knowledge and understanding of a range of materials, techniques and processes by selecting and using those that are appropriate to the requirements of the task. You should consider properties and working characteristics of materials and the processes used to manipulate them. In addition to this you should also be able to justify your selections by giving reasons for your choices.

In order to develop high-quality skills you must apply your knowledge and understanding of a range of materials, techniques and processes. It is likely that you will produce more than one practical outcome during this unit.

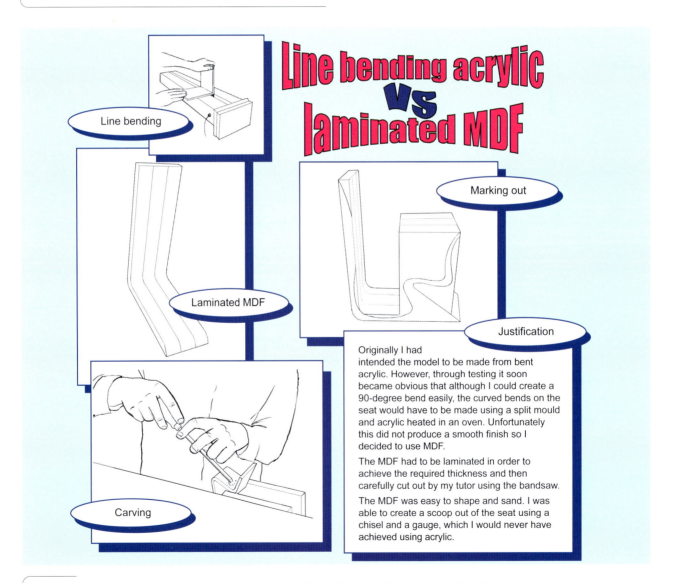

Line bending acrylic vs laminated MDF

Line bending

Marking out

Laminated MDF

Justification

Carving

Originally I had intended the model to be made from bent acrylic. However, through testing it soon became obvious that although I could create a 90-degree bend easily, the curved bends on the seat would have to be made using a split mould and acrylic heated in an oven. Unfortunately this did not produce a smooth finish so I decided to use MDF.

The MDF had to be laminated in order to achieve the required thickness and then carefully cut out by my tutor using the bandsaw.

The MDF was easy to shape and sand. I was able to create a scoop out of the seat using a chisel and a gauge, which I would never have achieved using acrylic.

Figure 1.16 Justification of materials and processes as part of a student's detailed evidence of a making activity.

The table below gives you some examples of possible projects and the wide variety of materials, techniques and processes open to you in your studies.

Table 1.1 Examples of projects including materials, techniques and processes covered.

Project	Materials covered	Techniques and processes covered
Perfume bottle and packaging	Bottle: cast acrylic block Packaging: cartonboard	Bottle: shaping and finishing acrylic on lathe, high-gloss finish on buffer, etc. Packaging: CAD/CAM using plotter/cutter to produce net; digital colour printing
Concept car	Car body: MDF Wheels: PVC tubing and foam	Car body: machine cutting, shaping using surforms, rasps and files, finishing using glasspaper and aerosol paints Wheels: hand cutting, shaping and gluing
Architectural modelling	Foamboard, acrylic sheet, polypropylene (PP) sheet, etc.	Cutting using craft knife scalpel, hot melt gluing, bending acrylic on line bender, etc.
Yoghurt pot	Yoghurt pot: MDF mould and high-impact polystyrene (HIPS) shell Lid: foil	Yoghurt pot: cutting, shaping and finishing MDF, vacuum forming and finishing Lid: dye sublimation printing onto foil
Mobile phone	Scale 2:1 Styrofoam™ model	Cutting on vibrosaw, shaping and finishing using files and glasspaper; hand-painting using acrylic paints

In order to achieve high marks in this section, you must show demanding and high-level making skills. Therefore, it is important that manufacturing tasks set by your teacher provide enough complexity and challenge to allow you to demonstrate your skill levels to the full. It is important to keep in mind, too, that the manufacturing tasks set in this unit should be designed to develop skills that you can call upon in your A2 coursework project. A single manufacturing project that involves a range of materials, processes and techniques that you can learn from can be as valid as two or three shorter but equally demanding exercises. However, by setting different exercises, the use of a range of materials, processes and techniques can be assured.

You will need to use a variety of skills and processes during your making activities, which may include computer-aided manufacture (CAM). Where this is a feature of your work, you should make sure that there is plenty of opportunity within the tasks to demonstrate other skills and competencies that you have gained through other making activities. While the use of CAM is to be encouraged, you must not over-use CAM. It is okay for you to dedicate one manufacturing exercise to the use of CAM in order to explore its capabilities, but you must offer evidence of other skills and techniques in other manufacturing exercises. Whenever CAM is used you must provide evidence of programming the computer numerical control (CNC) equipment. Where a mixture of CAM and other skills and techniques is used in a manufacturing exercise, CAM should not exceed 50 per cent of the work.

Throughout your making, you should be aware of the risks involved in using specific tools, equipment and processes and should take appropriate precautions to minimise those risks. A risk assessment of all relevant equipment is an appropriate way of recording this awareness.

Table 1.2 An example of a typical risk assessment for using a band facer in a school workshop.

Hazard	Risk	People at risk	Control measure
Abrasion – belt	Cuts/abrasions to hands	Teacher/student	1. Sliding guard, where fitted to be <6mm above work. 2. Pupils queuing to use machine must stand behind yellow line. 3. Never leave machine running unattended. 4. Isolate machine after use. 5. Staff use only after recognised training. 6. Student use after training and at teacher's discretion.
Trapping – gap between belt and table	Trapping of fingers in gap between belt and table	Teacher	1. Adjustments to be made by competent staff. Post-16 students at teacher discretion. 2. Power to be isolated during repair/maintenance.
Projectile – dust and debris	Dust and debris being propelled into eyes	Teacher/student	Safety spectacles provided in boxes next to machine
Dust – respiratory irritant	Effects of prolonged dust inhalation	Teacher	1. Dust extraction fitted and operational. 2. Dust extraction system serviced every year.
Dust – skin irritant	Effect of prolonged contact with wood dust	Teacher	Barrier cream available

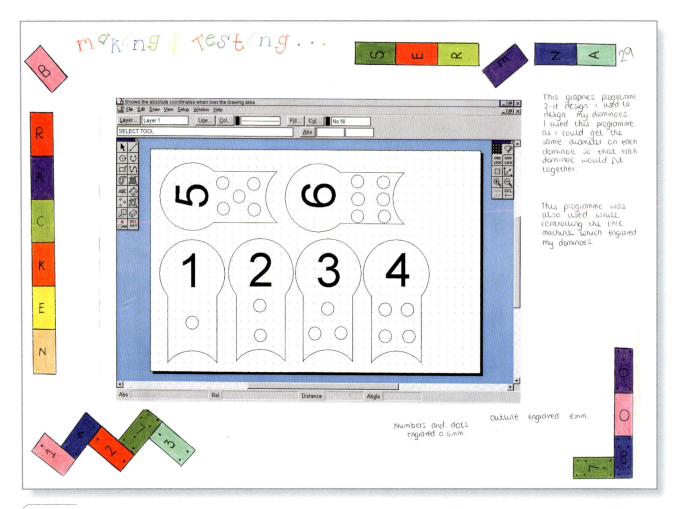

The following text appears within the figure area as handwritten annotations:

making & testing...

This graphics programme 2-d design I used to design my dominoes. I used this programme as I could get the same diameter on each domino so that each domino would fit together.

This programme was also used while controlling the CNC machine which engraved my dominoes.

outline engraved 8mm.

Numbers and dots engraved 0.5mm.

Figure 1.17 *CAM can be used to produce some components in a product. Here, the student has used CAM to produce multiple pieces for a young children's game.*

You will not be expected to produce a risk assessment for every piece of equipment you use. Annotated photographs of your making process should indicate where health and safety issues have arisen. For example, when using a band-facer your photograph should clearly show you with your safety goggles on with your hair tied back.

LINKS TO:

Unit 2: Health and safety describes the procedures for carrying out a risk assessment according to the Health and Safety Executive (HSE).

As proof of the quality of your making skills (and the level of demand of your work), photographs of your work must be evidenced to show that the product is complete, expertly made, well finished, etc. These photographs must clearly show any details of advanced skills, technical content, levels of difficulty and complexity of construction, so that you can achieve the marks you deserve. It is unlikely that a single photograph for each product will be enough to communicate all of the information required, so it will be better to take a series of photographs over a period of time during making. These should highlight the processes used and provide examples of precision and attention to detail that may not be otherwise noticed.

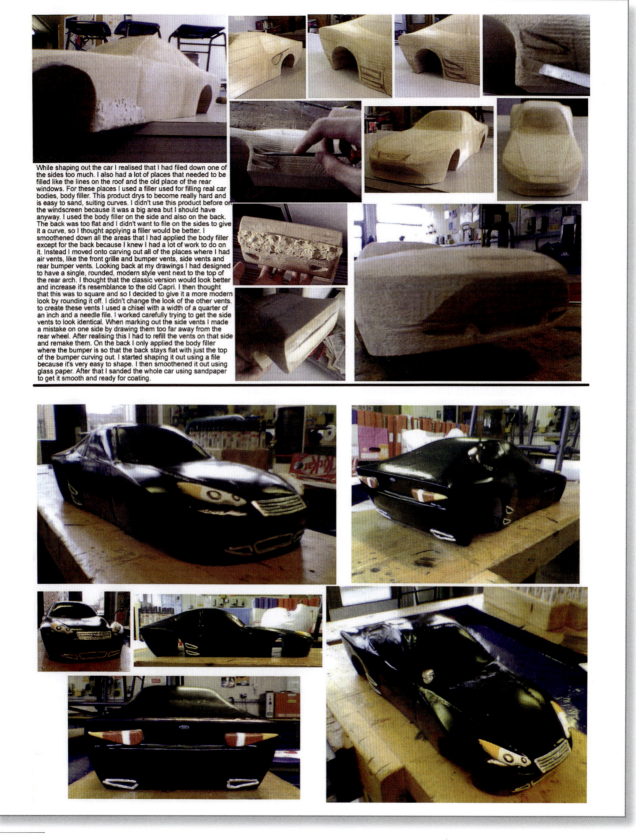

While shaping out the car I realised that I had filed down one of the sides too much. I also had a lot of places that needed to be filled like the lines on the roof and the old place of the rear windows. For these places I used a filler used for filling real car bodies, body filler. This product drys to become really hard and is easy to sand, suiting curves. I didn't use this product before on the windscreen because it was a big area but I should have anyway. I used the body filler on the side and also on the back. The back was too flat and I didn't want to file on the sides to give it a curve, so I thought applying a filler would be better. I smoothened down all the areas that I had applied the body filler except for the back because I knew I had a lot of work to do on it. Instead I moved onto carving out all of the places where I had air vents, like the front grille and bumper vents, side vents and rear bumper vents. Looking back at my drawings I had designed to have a single, rounded, modern style vent next to the top of the rear arch. I thought that the classic version would look better and increase it's resemblance to the old Capri. I then thought that this was to square and so I decided to give it a more modern look by rounding it off. I didn't change the look of the other vents. to create these vents I used a chisel with a width of a quarter of an inch and a needle file. I worked carefully trying to get the side vents to look identical. When marking out the side vents I made a mistake on one side by drawing them too far away from the rear wheel. After realising this I had to refill the vents on that side and remake them. On the back I only applied the body filler where the bumper is so that the back stays flat with just the top of the bumper curving out. I started shaping it out using a file because it's very easy to shape. I then smoothened it out using glass paper. After that I sanded the whole car using sandpaper to get it smooth and ready for coating.

Figure 1.18 *Part of a student's detailed photographic evidence of the making process.*

To be successful you will:

Assessment criteria: H. Making

Level of response	Mark range
Demonstrate a detailed understanding and justified selection of a range (1 mark) of appropriate materials (1 mark) and processes. (1 mark) Demonstrate demanding and high-quality making skills and techniques. (1 mark) Show accuracy and precision when working with a variety of materials, processes and techniques. (1 mark) Show high-level safety awareness that is evident throughout all aspects of manufacture. (1 mark)	13–18
Demonstrate a good understanding and selection of an appropriate range (1 mark) of materials (1 mark) and processes. (1 mark) Demonstrate competent making skills and techniques appropriate to a variety of materials and processes. (1 mark) Show attention to detail and some precision. (1 mark) Demonstrate an awareness of safe working practices for most specific skills and processes. (1 mark)	7–12
Demonstrate a limited understanding and selection of a narrow range (1 mark) of materials (1 mark) and processes. (1 mark) Use limited making skills and techniques. (1 mark) Demonstrate little attention to detail. (1 mark) Demonstrate an awareness of specific safe working practices during product manufacture. (1 mark)	1–6

I Testing (6 marks)

After making each product, you should then carry out tests to check their fitness for purpose against the set design requirements. Your finished product, as far as possible, should be tested under realistic conditions to determine its success and to check its performance and quality. You should describe in detail any testing carried out and justify this by stating what aspects you are testing and why you are doing so. Tests should be carried out objectively, and it would be beneficial to involve potential users so that you can receive reliable and unbiased third-party feedback. Well-annotated photographic evidence is a very good tool to use when describing the testing process.

To be successful you will:

Assessment criteria: I. Testing

Level of response	Mark range
Describe and justify a range of tests carried out to check the performance or quality of the product(s). (1 mark) Objectively reference relevant, measurable points of the design brief(s)/need(s). (1 mark) Use third-party testing. (1 mark)	4–6
Carry out one or more simple tests to check the performance or quality of the final product(s). (1 mark) Reference superficially some points of the design brief(s)/need(s). (1 mark) Record test results that are subjective. (1 mark)	1–3

Presentation of portfolio

Your portfolio of creative skills must be organised into **three** distinct sections clearly headed: product investigation, product design and product manufacture. It is important that each individual piece of work provided for assessment is evidenced in the appropriate section. This will allow your teacher to easily mark your work and provide the Edexcel moderator with a clear indication of your skills and ability.

While there is no defined limit to the number of pages you should include, it is envisaged that all requirements of this unit can be achieved within 25–30 A3-size pages. You may choose to produce your product investigation in A3 or A4 format. You can also submit your work electronically for moderation provided it is saved in a format that can be easily opened and read on any computer system, i.e. a PDF document.

Authentication

It is extremely important that you sign the authentication statement in your Candidate Assessment Booklet (CAB) before your work is marked. If you do not authenticate your work Edexcel will give you zero credit for this unit.

Testing: third-party feedback

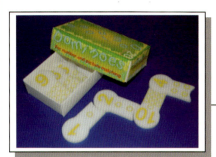

The box can be carried and dispensed very easily.

Joe found this box of dominoes very eye catching.
The colours are vivid and bold, which meets the most important point in my specification that the packaging advertises and appeals to the customers through the use of bright colours.

While Joe was playing with these dominoes he was thinking about getting it right.
This helps teach children how to count, which again meets my specification.

Joe enjoyed playing with the dominoes very much.
Asking his mother questions helps me to understand what the parents of young children want to buy.
She commented upon the colours – in her opinion each number could be a different colour as this would be more interesting and develop colour visualisation.

Figure 1.19 *Part of this student's testing involved third-party feedback provided by a member of the target market group (TMG).*

Design and Technology in Practice

Summary of expectations

1 What to expect

In this unit, you will develop your knowledge and understanding of a wide range of materials and processes used in the field of design and technology. It is important for you, as designers, to learn about materials and processes so that you can develop a greater understanding of how products can be designed and manufactured. You will also learn about industrial and commercial practices, the importance of quality checks and the health and safety issues that have to be considered at all times.

The knowledge and understanding that you develop in this unit can easily be applied to your *Unit 1: Portfolio of Creative Skills*.

2 How will it be assessed?

Your knowledge and understanding of topics in this unit will be externally assessed through a 1-hour 30-minute examination paper set and marked by Edexcel. The exam paper will be in the form of a question and answer booklet consisting of short-answer and extended-writing type questions.

The total number of marks for the paper is 70.

3 What will be assessed?

This unit is divided into four main sections, with each section outlining the specific knowledge and understanding that you need to learn.

2.1 Materials and components
- Materials.
- Components.

2.2 Industrial and commercial practice
- Scale of production.
- Graphical communication.
- Computer-generated graphics.
- Modelling and prototyping.
- Joining techniques.
- Industrial and commercial processes.
- Forming techniques.
- Finishing processes.
- Printing processes.

2.3 Quality
- Quality assurance systems and quality control in production.
- Quality standards.

2.4 Health and safety
- Health and Safety at Work Act (1974).

4 How to be successful in this unit

To be successful in this unit you will need to:
- have a clear understanding of the topics covered in this unit
- apply your knowledge and understanding to a given situation or context
- use specialist technical terminology where appropriate
- write clear and well-structured answers to the exam questions that target the amount of marks available.

5 How much is it worth?

This unit is worth 40 per cent of the AS level and 20 per cent of the overall full Advanced GCE.

Unit 2	Weighting
AS level	40%
Full GCE	20%

Materials and components

Getting started!

As a designer, you need to know about the properties of a wide range of materials and components so you can make informed choices about their use in certain products. What is the best material for a bike frame, for instance – aluminium or carbon fibre? Would the sales of a new perfume increase if it were packaged in a trendy acrylic tube rather than a traditional cartonboard box? Designers have to make these kinds of decisions with every product they design.

FACTFILE:

In this section you will need to develop knowledge and understanding, where appropriate, of the aesthetic, functional and mechanical properties of materials:

Property	Definition
Aesthetic properties	The visual and tactile qualities of a material.
Functional properties	The qualities a material must possess in order to be fit for purpose e.g. the correct weight, grade, size, etc.
Mechanical properties	A material's reaction to physical forces e.g. strength, plasticity, ductility, hardness, brittleness, malleability, etc.

1 Materials

Paper and board

The production of wood pulp

Paper and board are produced primarily from hard- and softwoods, although other materials can be used such as cotton, straw and hemp, producing papers with different properties. Wood is made up of cellulose fibres that are bound together by a material called lignin. In order to produce paper, these fibres must be separated from one another to form a mass of individual fibres called wood pulp. Softwood fibres are longer, offering greater strength, whereas hardwood fibres are shorter, offering a smoother, opaque finish. Wood pulp is produced by three basic methods: mechanical, chemical or waste.

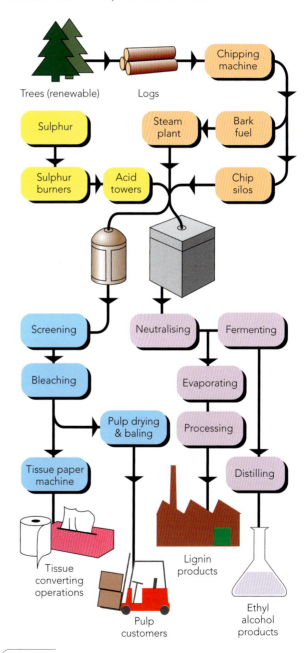

Figure 2.1 *The mechanical and chemical production of wood pulp.*

Mechanical pulp – The logs of coniferous trees are saturated with water and de-barked. The wood is ground down, which softens the lignin, and the mechanised forces separate the fibres to form 'groundwood pulp'. This pulp is screened to accept 1–2mm pieces, with larger pieces being re-circulated for additional screening. The resulting pulp can only be used for low-grade paper such as newspaper, so the pulp is bleached with peroxide or sodium hydroxide. This is the most widely used method in the UK for producing wood pulp.

Chemical pulp – After de-barking, the hard- and softwood logs are cut into 2cm chips along the grain. These are pounded into fragments and screened. The resulting pulp is stored and treated with either an acid or an alkali to break down the lignin. Most chemical pulp is made by the alkaline kraft process, or sulphate process, which uses caustic soda and sodium sulphate to 'cook' the wood pulp. The amount of fibre produced is lower than with mechanical methods, but the fibres are longer, stronger and contain fewer impurities.

Waste pulp – Recycled paper and board used for waste pulp is often used for lower grades of paper, as its strength, durability and colour are not as good as virgin fibres (produced using mechanical and chemical methods). Waste pulp is often mixed with virgin fibres to produce better quality papers as fibres become shorter and weaker and lose their papermaking qualities.

Manufacturers blend a variety of pulps and process them along with bonding agents and coloured pigments to produce paper with different qualities. These processes help to achieve a consistent colour and binding of fibres to create a better surface finish. A substance known as a sizing agent can also be added to improve water resistance, preventing ink from 'feathering' or bleeding on the surface.

Manufacturing paper using the Fourdrinier process

Modern paper production is achieved using a Fourdrinier machine. The Fourdrinier transforms wood pulp into a final paper product through four main sections.

- The **Wet end** starts with the wood pulp diluted to 99 per cent water and 1 per cent fibre to form a slurry that is held in the head box. A continuous stream of slurry is pumped from the head box through an adjustable slit (called the *slice*) onto a moving gauze wire belt that vibrates to drain off some of the water and allow the fibres to interweave. Raised patterns formed in the gauze create the watermark – a feature in many high-quality writing papers.

Table 2.1 Advantages and disadvantages of methods of producing wood pulp.

Method	Advantages	Disadvantages
Mechanical pulping	• Provides a 90% yield from the pulpwood as it uses the whole of the log except for the bark. • Investment costs for mechanical pulp mills are relatively low in comparison with other types of mill. • Well suited for 'bulk' grades of paper, i.e. newsprint and packaging boards. • Can be bleached to produce higher value-added products.	• Lower strength characteristics than softwood chemical pulps. • Paper can 'yellow' when exposed to bright lights due to high lignin content.
Chemical pulping	• Higher quality wood pulp produced with longer, stronger fibres that contain fewer impurities. • Produces 'chlorine-free' disposable products. • Waste lignin from the process can be burnt as a fuel oil substitute, often supplying power to the national grid or steam to local domestic heating plants.	• Lower yield than mechanical methods as the lignin is completely dissolved and separated from the fibres. • No chemical pulp is produced in the UK, therefore it must be imported.
Waste pulping	• Makes use of recycled papers, which are a sustainable resource. • Well suited for 'bulk' grades of paper, i.e. newsprint, tissue and packaging boards.	• Cannot be recycled indefinitely as pulp loses quality – virgin pulp needs to be added. • Does not save any more energy in processing than other methods. • Requires considerable processing and additives to produce good-quality paper.

- The **Press section** uses a system of nip presses or rollers that wrings out the majority of excess water from the pulp and stretches it out into a rough paper. It is at this stage that the thickness of the paper/board is determined. The gap between the press rollers is adjusted to allow for differing thicknesses such as card. The term card usually refers to paper which has a density greater than 160gsm (grams per square metre).

- The **Dryer section** dries the paper using a series of steam-heated rollers by removing the moisture (just like ironing clothes). The resulting paper has a water content of 4–6 per cent and sizing agents, starches and resins can be added to enhance the paper's properties.

- The **Calendar section** comprises a series of rollers through which the paper is fed in order to smooth it out and give it a uniform thickness. The pressure applied to the paper by these rollers determines the finish of the paper.

The paper is then wound onto a roll after calendering (known as a web) and stored. It can be placed onto a precision cutting machine to produce the desired size or shipped to a printer for web-fed printing.

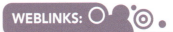

WEBLINKS:

www.ppic.org.uk – Pulp and paper information centre

Properties of paper and board

The choice of paper is essential in how printed materials are presented. Choosing the most appropriate paper for its intended application is a combination of personal preference and discussion with the client in order to determine the look of the end product. In general, the correct choice of paper must satisfy:

- the design requirements of the client brief, e.g. durability, surface finish, colour, texture, opacity, weight and size

- the demands of the printing process or surface decoration, e.g. will the printing inks used for offset lithography provide a quality finish on the paper?

- economic considerations, e.g. scale of production.

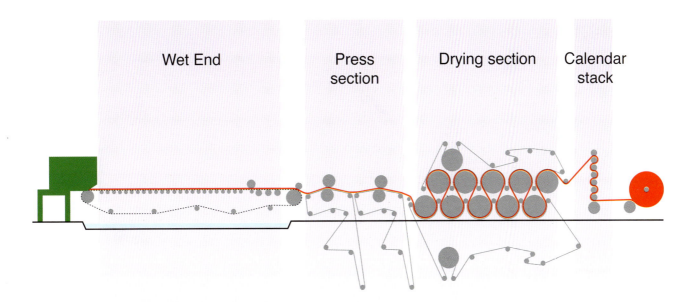

Figure 2.2 The Fourdrinier papermaking process.

Weight and size

Paper is defined in weight as gsm, with 80gsm being the weight of average copier paper. Card and board are measured in micrometres (microns). A paper is usually classified as a board when it is greater than 220gsm and more often than not is made from more than one ply. The thickness of card and board can be gauged by the number of ply it consists of, and is measured in microns (1000ths of a millimetre).

Paper, card and board are most commonly available in metric 'A' sizes (i.e. A5 through to A0). However, 'B' sizes and old imperial measurements are also widely used.

Table 2.2 *Common thicknesses of paper and board.*

Number of ply	Microns
2	200
3	230
4	280
6	360
8	500
10	580
12	750

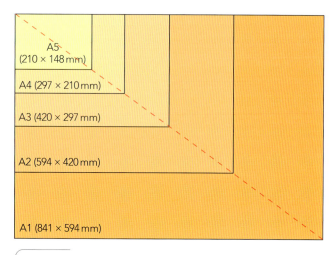

Figure 2.3 *Common 'A' sizes of paper and board.*

A5 (210 × 148 mm)
A4 (297 × 210 mm)
A3 (420 × 297 mm)
A2 (594 × 420 mm)
A1 (841 × 594 mm)

Table 2.3 *Common drawing papers, commercial printing papers and commercial card and board.*

Type	Weight	Description	Applications	Properties	Cost
Common drawing papers:					
Layout paper	Around 50gsm	Thin translucent paper with a smooth surface.	Outline sketches of proposed page layouts. Sketching and developing ideas. Marker renderings.	Translucent property allows tracing through onto another sheet. Accepts most drawing media (except paints).	Relatively expensive.
Tracing paper	60–90gsm	Thin transparent paper with a smooth surface. Pale grey in appearance.	Same as layout paper. Heavier weight preferred by draughtsmen.	Allows tracing through onto another sheet in order to develop design ideas.	Heavier weight can be quite expensive.
Copier paper	80gsm	Lightweight grade of quality paper. Good quality bleached surface.	B/W photocopying and printing from inkjet and laser printers. Smooth finish for colour printing. General use for sketching and writing.	Bright white and available in a range of colours.	Inexpensive when purchased in bulk.
Cartridge paper	120–150gsm	Creamy-white paper. Smooth surface with a slight texture.	Good general purpose drawing paper. Heavier weights can be used with paints.	Completely opaque. Accepts most drawing media.	More expensive than copier paper.
Commercial printing papers:					
Bond paper	Greater than 50gsm	High-quality durable writing paper that often carries a watermark. The name comes from having originally been made for documents such as government bonds.	Letterheads and other stationery and as paper for electronic printers. Widely employed for graphic work involving pencil, pen and felt-tip marker.	Largely made from rag pulp, which produces a stronger paper than wood pulp. Long-wearing paper. Available in a wide range of colours.	Inexpensive when purchased in bulk.

Coated	70–300gsm	Covered with a suspension of china clay, pigment and adhesive to give a smooth surface finish.	Wide range of high-quality print jobs.	The coating levels the minute pits between the fibres in the base paper, giving a smooth flat surface for printing. Range of finishes available from high gloss, matt, satin and silk.	Relatively expensive.
Commercial card and board:					
Recycled card	Greater than 220gsm	Card made from varying percentages of recovered waste pulp.	'Environmentally friendly' products, disposable items such as trays and inner packaging, sleeves for take-away hot drinks.	Lower quality than cards made with virgin pulp. Flecked appearance can have a quality of its own. Can be coloured for visual effect.	Inexpensive.
Mounting board	1000–1500 microns	Relatively thick board with colour on one side only (white on back).	Mounting work for presentations and displays. Work can be mounted flat or behind a frame mounting.	Very high quality, strong and rigid board. Available in a range of colours (wide range of pastel colours).	Expensive.

Cartonboards

Cartonboards are used extensively in the retail packaging industry where specific properties are required. These boards must be suitable for high-quality, high-speed printing and for cutting, creasing and gluing using very-high-speed automated packaging equipment.

Advantages of using cartonboard include:
- total graphic coverage and excellent print quality
- excellent protection in structural packaging nets

- relatively inexpensive to produce and process
- can be recycled.

LINKS TO:

Unit 3: Industrial and commercial practice: Industrial and commercial practices of creating structural packaging nets.

Table 2.4 Common cartonboards.

Board	Composition	Applications	Properties	Cost
Folding boxboard	Usually consists of a bleached virgin pulp top surface, unbleached pulp middle layers and a bleached pulp inside layer.	Widely used for the majority of food packaging and for all general carton applications.	• Excellent for scoring, bending and creasing without splitting. • Excellent printing surface.	Relatively inexpensive
Corrugated board	Constructed from a fluted paper layer sandwiched between two paper liners.	Protective packaging for fragile goods. The most commonly used box-making material.	• Excellent impact resistance. • Has excellent strength for its weight. • Low cost. • Recyclable.	Relatively inexpensive
Solid white board	Made entirely from pure bleached wood pulp.	Packaging for frozen foods, ice-cream, pharmaceuticals and cosmetics.	• Very strong and rigid. • Excellent printing surface.	Expensive
Foil-lined board	Consists of a laminated foil coating (can be used on all of the above boards). Foil available in matt or gloss finish and in silver or gold colours.	Cosmetic cartons, pre-packed food packages.	• Very strong visual impact. • Foil provides an excellent barrier against moisture.	Expensive

Metals

Metals can be divided into three main categories.

- **Ferrous metals** – which contain mainly ferrite or iron i.e. steel. It also includes those with small additions of other substances e.g. carbon steels. Almost all are magnetic.

- **Non-ferrous metals** – which contain no iron e.g. aluminium and tin. They are not magnetic.

- **Alloys** – which are formed by mixing two or more metals and, on occasion, other elements to produce metals with enhanced properties e.g. stainless steel (steel alloy) and duralumin (aluminium alloy).

Metals in commercial packaging

Metals are commonly used in a wide range of commercial packaging applications, from the standard components of drinks cans and bottle closures to special promotional containers that suggest quality. There are several advantages of using metals for commercial packaging.

- Added security, as sealed cans cannot be tampered with without obvious visible signs.

- Containers can be made in a variety of standard sizes and shapes e.g. drinks cans, and custom-made styles e.g. chocolate and biscuit tins.

- Containers can be embossed or de-bossed to provide surface textures and visual appeal.

- Metals can be directly printed on to or a paper label added for total graphic coverage, providing an effective point-of-sale display.

Figure 2.4 Aluminium and steel (sometimes tin-plated) are the only two metals used in commercial packaging.

THINK ABOUT THIS!

Many companies package their products in metal containers for special occasions such as Christmas or Easter. Why do companies do this? For example, a shortbread manufacturer uses a cartonboard package for the remainder of the year but at Christmas puts them in a tin.

Aluminium – Aluminium is a strong but lightweight metal that is easily moulded into a variety of containers. It is a non-ferrous metal so it does not rust, which makes it perfect for making a range of containers. Three quarters of drinks cans are made from aluminium and about 15 per cent of aerosol cans are aluminium.

Aluminium is a pure metal and is a naturally occurring element that is mined from beneath the land and sea. It is the most plentiful metal element in the Earth's crust and is produced from the ore bauxite. The extraction of alumina from the bauxite ore, and subsequent production of aluminium from alumina, uses a lot of energy. However, aluminium is easily recycled as it can be melted down over and over again without being spoiled and can be made into another drinks can. Recycling aluminium is much more sustainable than producing it from alumina, because it saves considerable amounts of energy. For example, recycling one aluminium can saves enough energy to run a TV for 3 hours and if all the aluminium cans in Britain were recycled, about 14 million fewer dustbins would be filled every year. Currently, just 34 per cent of aluminium drinks cans are recycled.

Tin – Tin is obtained from the mineral cassiterite, which is tin oxide. Cassiterite is smelted to a metal by reduction with carbon, most commonly in a reverberatory furnace. However, this metal still contains impurities, so it is necessary to refine the tin to make it commercially useful. One of the most important properties of tin is the ease with which it alloys or mixes with the majority of other metals. It is this quality together with the low melting point that makes it an essential ingredient of most solders for attaching electrical components to circuit boards. It is not toxic and it does not corrode very rapidly making it ideal as protection for steel in food and drinks 'tin cans'.

Steel – In the food packaging industry, steel is usually in the form of tinplate, which consists of cold rolled steel sheet coated with a thin layer of tin to prevent corrosion. Therefore, steel cans are often called 'tin cans'. When steel is rolled into thin sheets it is a lightweight material that is easily moulded and is used to make a range of food cans and most aerosol cans.

Steel is produced from iron ore, which is widely found and mined. It takes a lot of energy to turn iron ore into steel, as it has to be heated in a huge furnace to very high temperatures to separate it from the other materials present. However, steel can be easily recycled. It takes about 75 per cent less energy to recycle steel than to make it from iron ore. Steel is a ferrous metal, and therefore magnetic, making it easy to sort using powerful electromagnets at recycling plants that separate the steel from other metals. Currently, around 51 per cent of all steel packaging is recycled.

WEBLINKS:

www.mpma.org.uk – Metal Packaging Manufacturers Association, offering information on all aspects of metals in packaging

Figure 2.5 *The Art Deco Chrysler Building in New York is clad with stainless steel for its aesthetic and functional properties.*

FACTFILE:

Alloy	Composition		Properties	Applications
Stainless steel	Carbon steel (87%)	Strength and rigidity	Advantages: • excellent resistance to corrosion, oxidation (rust) and staining • high-quality surface finish • low maintenance • 100% recyclable. Disadvantages: • relatively expensive • relatively hard to machine.	With over 150 grades available (increasing the chromium content up to 26% for harsh environments), there are a wide range of applications, including: cutlery, jewellery, cookware, sterile surgical instruments, building construction, etc.
	Chromium (13%)	Resistance to wear and corrosion		
Duralumin	Aluminium (93.5%)	Strength and lightness	Advantages: • excellent strength to weight ratio • extremely hard and tough • highly resistant to stress-corrosion cracking • machines and casts well • high performance in extreme temperatures. Disadvantages: • can suffer from corrosion • becomes brittle through fatigue.	Applications make use of its excellent strength to weight ratio and casting ability, including: aircraft and vehicle structures (shells and frames), precision tools and components including bicycle and engine parts, etc.
	Copper (4.4%)	Strength		
	Magnesium (1.5%)	Alloying agent (strength and lightness)		
	Manganese (0.6%)	Deoxidising (removing oxygen)		

Polymers

Polymers in commercial packaging

Thermoplastic polymers are widely used in commercial packaging. A thermoplastic is a material that, once heated, can be formed into a variety of interesting shapes using different forming techniques such as blow moulding. Once cooled, the shape remains permanent. However, thermoplastics can be re-heated, softened, shaped and cooled many times over, so they can be recovered and recycled easily. The main advantages of using thermoplastics in commercial packaging are that they are:

- lightweight and versatile
- strong, tough, rigid, durable, and impact and water resistant
- easily formed and moulded
- easily printed on
- inexpensive
- recyclable.

Polymers used in packaging can be identified by an internationally recognised coding system moulded into the base of the package or printed on the label. This system enables polymers to be easily identified and sorted prior to recycling. Each polymer has its own useful properties that make it suitable for use in different areas of the packaging industry.

Figure 2.6 *Polymers have proved to be an extremely versatile material for packaging a wide range of products.*

LINKS TO:

Unit 3: Industrial and commercial practice: Forming techniques for blow moulding, injection moulding and vacuum forming thermoplastics.

Styrofoam™ for block modelling

Styrofoam is a blue or pink extruded polystyrene foam manufactured for the construction industry for insulating buildings. However, it makes an excellent modelling material, especially for block modelling purposes. The process of extruding foamed polystyrene results in a material with uniformly small, closed cells, which gives it all of the properties that a modelling material should include: great rigidity, high compressive strength, easily cut and shaped with normal hand tools, and sanded to a smooth surface finish. Styrofoam is also available in a range of

thicknesses, can be glued together with polyvinyl acetate (PVA) to create larger block sizes and painted with acrylic paints to give a good-quality finish. A higher-quality finish can be achieved by coating the Styrofoam in layers of plaster, sanding smooth, and then applying a surface finish using spray paints. Whilst Styrofoam is a good modelling material it has a number of disadvantages. It is weak and the surface breaks away very easily and therefore it is unsuitable for models that require any fine detail. Also, the surface can become 'dinted' when using a file or ripped when sanding if care is not taken.

LINKS TO:

Unit 2: Industrial and commercial practice: Modelling and prototyping using block modelling to aid the development of products.

Table 2.5 *Common thermoplastic polymers used in packaging.*

Thermoplastic	ID code	Properties	Applications
PET Polyethylene terephthalate	1 PET	• Excellent barrier against atmospheric gases. • Prevents gas from escaping package. • Does not flavour the contents. • Sparkling 'crystal clear' appearance. • Very tough. • Lightweight – low density.	Carbonated (fizzy drinks) bottles. Packaging for highly flavoured food. Microwavable food trays.
HDPE High density polyethylene	2 HDPE	• Highly resistant to chemicals. • Good barrier to water. • Tough and hard wearing. • Decorative when coloured. • Lightweight and floats on water. • Rigid.	Unbreakable bottles (for washing-up liquid, detergents, cosmetics, toiletries, etc.). Very thin packaging sheets.
PVC Polyvinyl chloride	3 PVC	• Weather resistant – does not rot. • Chemical resistant – does not corrode. • Protects products from moisture and gases while holding-in preserving gases. • Strong, good abrasive resistance and tough. • Can be manufactured either rigid or flexible.	Packaging for toiletries, pharmaceutical products, food and confectionery, water and fruit juices.
LDPE Low density polyethylene	4 LDPE	• Good resistance to chemicals. • Good barrier to water, but not gases. • Tough and hard wearing. • Decorative when coloured. • Very light and floats on water. • Very flexible.	Stretch wrapping (cling film), milk carton coatings.
PP Polypropylene	5 PP	• Lightweight. • Rigid. • Excellent chemical resistance. • Versatile – can be stiffer than polyethylene or very flexible. • Low moisture absorption. • Good impact resistance.	Food packaging – yoghurt and margarine pots, sweet and snack wrappers. Used for laminating paper and board.
PS Polystyrene	6 PS	Rigid polystyrene: • transparent • rigid • lightweight • low water absorption. Expanded polystyrene: • excellent impact resistance • very good heat insulation • durable • lightweight • low water absorption.	Rigid: Food packaging, e.g. yoghurt pots, CD jewel cases, audio cassette cases. Expanded: Egg cartons, fruit, vegetable and meat trays, cups, etc. Packing for electrical and fragile products.

Figure 2.7 *Acrylic is often used for shop signage.*

Acrylic

Acrylic is the common name for polymethyl methacrylate (PMMA) and sometimes referred to by its tradenames Plexiglas® or Perspex®. Acrylic is usually cast into sheets but is also available in rods and tubes and is self-finishing, so it does not require an applied surface finish. It has a wide range of applications within graphic products from point-of-sale displays to shop signage. Architectural-grade acrylic, for example, has to satisfy several high performance requirements when used in shop signage. It has to be able to withstand extreme weather conditions, be chemical resistant, be durable by resisting long-term stresses, be easy to fabricate and have excellent aesthetic properties. Clear or frosted acrylic is often used in point-of-sale displays and illuminated signage as a low-cost and lightweight substitute to glass. The main disadvantages of acrylic are its brittleness and its low scratch resistance. Also, cracks easily form and spread through acrylic. Because acrylic scratches easily and can snap, great care must be taken when using it.

WEBLINKS:

www.lucitesolutions.com – Manufacturers of Perspex® and related products

www.plexiglas.com – Manufacturers of Plexiglas® and related products

Environmental concerns

The main disadvantage of the widespread use of polymers seems to be the ongoing environmental concern of how sustainable it is. In the first instance, oil is the raw material of synthetic polymers, and is not only a finite resource but consumes a lot of energy in the production process and produces pollution. The manufacture of packaging, blow moulding for example, also uses a lot of energy. Then comes the problem of disposal. Polymers are durable and degrade very slowly, which is a problem for landfill sites. Incineration may not be the answer either, as in some cases burning polymers can release toxic fumes.

Recycling may be the answer to these environmental concerns as thermoplastics can be moulded into new products. The main problems at present are collection and sorting. Many councils operate recycling collection schemes for polymers, glass, metal, paper and organic waste; however, polymers are the hardest to recycle. They can be identified and sorted by means of their ID code but this is a slow and labour-intensive process. There are mechanical methods of sorting but these are not widespread. There is also the problem of contamination, as the majority of bottles, for example, are made of one polymer while the closure is made of another. Recycling certain types of polymers can also be unprofitable: for example, polystyrene is rarely recycled because it is usually not cost effective.

LINKS TO:

You will be studying environmental concerns in **Unit 3: Sustainability** in a lot more detail at A2 level.

Woods for modelling

Hardwoods

Hardwoods are produced from broad-leaved trees whose seeds are enclosed. Examples include oak, mahogany, beech, ash and elm. Hardwood trees commonly grow in warmer climates such as Africa and South America and take about 100 years to reach maturity. They are usually tough and strong and because of their close grain provide highly decorative surface finishes. Because of their age and location, many hardwoods are expensive to buy and may only be used for high-quality products. The exception is balsa wood, which has been used for modelling applications for many years as it is relatively inexpensive and easy to shape.

Figure 2.8 *A model aircraft made out of balsa wood.*

Softwoods

Softwoods are produced from cone-bearing conifers with needle-like leaves. Examples include Scots pine (red deal), Parana pine and whitewood. As softwoods grow more quickly than hardwoods (around 30 years) they can be forested and replanted, which means they are in abundance and therefore cheaper to buy. Softwoods are generally easier to work with and are lightweight, making them more suitable for modelling applications. However, as with all woods, they contain a grain that makes them harder to shape when block modelling. In this case, a composite material such as medium density fibreboard (MDF) is more applicable.

THINK ABOUT THIS!

There are many materials that you could use for making models of products for your Portfolio of Creative Skills. The key to success is to use a wide range of modelling materials instead of the usual Styrofoam™ and MDF so, for example, try to model something out of balsa for a change; it is available from most modelling and craft shops.

Composites

When two or more materials are combined by bonding, a composite material is formed. The resulting material has improved mechanical, functional and aesthetic properties and, as with most composites, it will have excellent strength to weight ratios.

Glass-reinforced plastics

Glass-reinforced plastic (GRP) composite is made of a polyester or epoxy resin reinforced by fine fibres of glass in the form of a woven matting. The plastic resins are strong in compressive strength but relatively weak in tensile strength, whereas the glass fibres are very strong in tension but have no compressive strength. By combining the two materials GRP becomes a material that has both compressive and tensile strength. The resin exists in a liquid form and has a catalyst or hardener added to become a solid. The glass fibre strands provide the basic structure while the resin

Table 2.6 *Hardwoods and softwoods used in the creation of models and prototypes.*

Wood	Type	Origin	Properties
Jelutong	Hardwood	Indonesia	• Low density. • Straight grain. • Fine texture.
Balsa	Hardwood	South America	• Extremely buoyant. • Very soft and light. • Low density with high strength.
Scots pine	Softwood	Northern Europe	• Easy to work. • Knotty and prone to warping.
Parana pine	Softwood	South America	• Hard, straight grain (often knot-free). • Fairly strong, durable and easy to work. • Smooth finish.
Whitewood	Softwood	Northern Europe, North America	• Fairly strong but not durable. • Easy to work. • Very resistant to splitting.

Figure 2.9 *GRP is used in the manufacture of many fairground rides and attractions.*

with its additives bonds the fibres together and provides a lightweight rigid structure. The two materials may be used uniformly or the glass may be specifically placed in those portions of the structure that will experience tensile loads. The extremely smooth GRP finish seen on boats and some cars is achieved by a combination of a highly polished surface on the mould used and careful application of the first layer, known as the gel coat. The glass matting which is laid on top of the gel coat to provide the basic structure leaves a very rough finish. Therefore, one surface of GRP is highly polished whilst the other is extremely rough.

Carbon fibre

More recently, carbon fibre has been commercially developed in a similar form to that of glass fibre. This carbon composite is made up of carbon fibres, which take tensile loads, set into a polymer resin matrix that takes the compressive loads. Carbon fibre is a filament material incorporating thousands of filaments that are woven to form a fabric. However, this fabric only has strength in tension. So the woven fabric is placed in different directions to cover tensile loads in all directions, while being supported by a rigid, compression bearing matrix of resin.

Carbon fibres are much stronger than GRP and are ideal for high-performance structural applications in aircraft,

Figure 2.10 *An excellent strength to weight ratio is one of the reasons why carbon fibre is used in preference to metals in high-performance racing bikes.*

sports equipment and F1 racing car manufacture. Carbon composites have unrivalled mechanical properties and, in most load-bearing applications where weight is an issue, will easily out-perform any metal alternative. For example, carbon fibre has more than four times the tensile strength of the best steel alloys, at just a quarter of the weight! It also has a much better fatigue life.

Medium-density fibreboard

One of the most widespread and commonly used composite materials is medium-density fibreboard (MDF). MDF is primarily made from wood waste (or specifically grown softwoods) in the form of wood chips, which are subjected to heat and pressure in order to soften the fibres and produce a fine, fluffy and lightweight pulp. This pulp is then mixed with a synthetic resin adhesive to bond the fibres and produce a uniform structure and heat pressed to form a fine textured surface. After pressing, the MDF is cooled, trimmed and sanded. In certain applications

boards are also laminated for extra strength. MDF can be worked like wood but with the added advantage that it has no grain to work with. It finishes well with a variety of surface treatments and is available with a veneered surface for decorative effect.

As is the case with all composites, there are some potential hazards involved in their use. As well as fumes from glues and resins, as a result of the very fine fibres great care must be taken when undertaking any form of cutting, drilling and especially sanding. Respiratory equipment should be used since the dust can cause irritation of the skin, throat and nasal passage and appropriate dust extraction should also be activated.

LINKS TO:

Unit 2: Health and Safety: HSE risk assessments for workshop practices.

Table 2.7 *Applications, advantages and disadvantages of composite materials.*

Composite	Applications	Advantages	Disadvantages
Glass reinforced plastics (GRP)	Rotor blades of wind turbines, canoes, fish ponds, vehicle bodies, kiddies sit 'n' ride and fairground rides, etc.	• Excellent strength to weight ratio. • Resistent to corrosion. • Water resistent. • Ideal for external shell structures. • Wide range of colours as pigments can be added to the resin. • Can be repaired easily.	• Expensive material. • Specialised manufacturing process required. • High-quality mould needed.
Carbon fibre	Sports equipment, e.g. tennis raquets and fishing rods, bicycle frames and wheels, aircraft and vehicle componets, etc.	• Excellent strength to weight ratio. • Better tensile strength than steel alloys. • Can be engineered to be anisotropic (fabric orientated in different directions to provide strength in specific areas of structure). • Can be formed into complex and aerodynamic one-piece structures (distribute stress efficiently).	• Very expensive material. • Only available in black. • Highly specialised manufacturing processes required. • Cannot be easily repaired as structure loses integrity. • Cannot be easily recycled.
Medium-density fibreboard (MDF)	Flat-pack furniture, general joinery work, moulds for forming processes, etc.	• Less expensive than natural timbers. • Available in large sheet sizes and range of thicknesses. • Isotropic (no grain), so no tendency to split. • Consistent strength in all directions.	• Heavier (the resins are heavy). • Requires appropriate finishes to seal surface fibres. • Swells and breaks when waterlogged. • Warps or expands if not sealed. • Contains urea-formaldehyde which may cause eye and lung irritation when cutting and sanding. • Dulls blades more quickly than many woods.

Modern materials and products

Liquid crystal displays

Liquid crystals are organic, carbon-based compounds that can exhibit both liquid and solid crystal characteristics. When a cell containing a liquid crystal has a voltage applied, and on which light falls, it appears to go 'dark'. This is caused by the molecular rearrangement within the liquid crystal. In the case of a digital clock or wrist-watch, a liquid crystal display (LCD) display has a pattern of conducting electrodes that is capable of displaying numbers via a seven-segment display. The numbers are made to appear on the LCD by applying a voltage to certain segments, which go dark in relation to the silvered background. As very small amounts of current are needed to power an LCD display they are ideal for portable electronic devices such as mobile phones, as battery life can be extended.

With the rapid advance of LCD technology came the full-colour LCD display commonly used in laptops. In this case, each pixel is divided up into three sub-pixels with red, green or blue filters. By controlling and varying the voltage applied, the intensity of each sub-pixel can range over 256 colours. LCDs have now evolved considerably and are at the forefront of modern domestic appliance technology with ever flatter, high-resolution LCD televisions and computer screens.

WEBLINKS:

www.howstuffworks.com – Full details on how LCDs work

Phosphorescent pigments

Phosphorescent pigments are often referred to as 'afterglow' materials. These materials absorb energy or 'charge' in normal daytime conditions and are capable of storing that energy for some time, then releasing it relatively slowly in the form of light. They are non-radioactive like the old forms of afterglow materials, where a polycrystalline inorganic zinc sulphide pigment can be used to create a green afterglow or alkaline earth sulphides for a red or blue afterglow. Phosphorescent pigments can be manufactured into polymers used in paints or inks to produce products like safety signage and watch faces that can be seen in complete darkness.

Light (normal conditions)

Dark (power failure)

Figure 2.11 *The use of phosphorescent pigments in safety signage.*

WEBLINKS:

www.dualglo.com – Phosphorescent materials and products

Electroluminescent lighting

Electroluminescent (EL) lighting converts electrical energy into light (or luminescence) by applying a voltage across electrodes. An organic phosphor is sandwiched between two conductors and, as the electric current is applied, it rapidly charges phosphor crystals, which emit radiation in the form of visible light. EL lighting has extremely low power consumption, which makes it ideal for use as a backlight for LCD displays; for example, the blue glow on a digital wrist watch. It is used to great effect in advertising to produce displays and posters with high visual impact. EL lighting is produced in the form of paper-thin wires, strips or panels, which are applied to designs that can then be animated by lighting up different areas of the poster. As they are so thin it is possible to utilise all the traditional methods of advertising, such as bus shelters, sides of buses and billboards as they are waterproof, highly visible and extremely reliable.

WEBLINKS:

www.edmonds.co.uk/electroluminescent-displays – EL displays and posters in action

Figure 2.12 *The distinctive blue glow of electroluminescent backlighting for LCDs.*

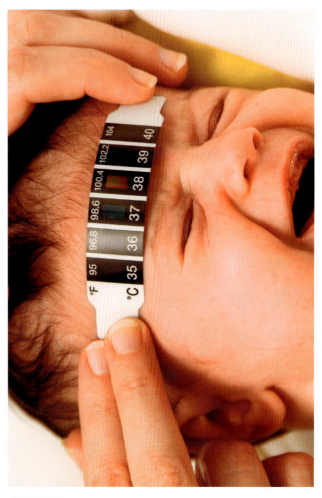

Figure 2.13 *This forehead thermometer makes use of thermochromic liquid crystals.*

Smart materials

Smart materials have been developed through the invention of new or improved technologies, for example as a result of manufactured materials or human intervention. Smart materials respond to differences in temperature or light (and other inputs such as pressure and sound) and change in some way as a result. They are classified as smart because they sense the conditions in their environment and respond to those conditions. Smart materials appear to 'think' and some have a 'memory' as they revert back to their original state as soon as the input is removed (unlike thermoplastics).

Thermochromic liquid crystals

Thermochromic liquid crystals are used in a number of applications, including forehead thermometers, battery test panels and special printing effects for promotional items. In the case of a forehead thermometer, a layer of conductive ink is screen printed on to the reverse of the thermometer strip – this area makes contact with the forehead. On top of the conductive ink is a layer of normal ink that conveys the temperature gauge colour bars. Finally, there is the thermochromic layer, which is black when cool. By pressing the thermometer to the forehead, the temperature generated turns the thermochromic ink translucent. This reveals the temperature gauge colour bars, which are printed in normal ink. Depending on inner body temperature, most or all of the thermochromic ink will heat to the temperature needed to become translucent. The same process applies to battery test panels, where the electrical charge of the battery generates the heat required.

Other special printing inks are available to enhance printed materials specially for promotional use. Thermoreactive inks can be used to reveal graphics if a warm hand is placed over them or, conversely, if they are placed in a fridge – an ideal device for revealing a lucky winner on a promotional pack. As well as inks that react to changes in temperature, some inks react to ultraviolet (UV) radiation in natural sunlight. These are known as photochromic inks.

Piezoelectric crystals

A piezoelectric crystal is a material that expands and contracts when electric current is applied. The piezoelectric effect converts this mechanical stress or vibration into electrical signals and vice versa. Computer manufacturer Epson, for example, uses piezoelectric crystal technology in its inkjet printers. A piezoelectric crystal is located at the back of the ink reservoir of each nozzle. The crystal receives

Figure 2.14 Piezoelectric crystals are used in music modules for musical greetings cards.

a tiny electric charge that causes it to vibrate. When the crystal vibrates inward, it forces a tiny amount of ink out of the nozzle. When it vibrates out, it pulls some more ink into the reservoir to replace the ink sprayed out. Piezoelectric crystals are used in a wide range of products and systems as either actuators or transducers.

Table 2.8 Piezoelectric crystals as actuators and transducers.

Type	Example	How it works
Actuator A device for controlling a mechanism or system	Musical greetings card	The music module is activated by opening the card, which removes an insulating tongue from between a pair of switch contacts on the module. The piezoelectric crystal acts as a tiny speaker driver allowing it to generate a pre-programmed sound stored on an integrated circuit mounted on the module.
Transducer A device that converts a signal from one form to another	Electronic drum kit	When an electronic drum pad is struck, a voltage change is triggered in the embedded piezoelectric transducer. The resultant signals are translated into digital waveforms, which produce the desired percussion sound assigned to that particular trigger pad.

Smart ink

Smart ink, also known as electronic ink or electronic paper, is a display technology designed to mimic the appearance of ordinary ink on paper. Electronic paper was developed in order to overcome some of the limitations of computer monitors. For example, the backlighting of monitors is hard on the human eye, whereas electronic paper reflects light just like normal paper. It is easier to read at an angle than flatscreen monitors. It is lightweight, durable, and highly flexible compared with other display technologies, though it is not as flexible as paper.

Smart ink is currently being developed for applications such as electronic books, capable of storing digital versions of many books. A major advantage of smart ink is that the pixels have an inherently stable 'memory effect' that requires no power to maintain an image. Displays only draw on battery power when text is refreshed, which means they can display about 10,000 pages before the batteries need changing. Further technological developments will include electronic newspapers where headlines can be constantly updated and animated images and video clips used.

How smart ink works

Each pixel point on the display is a tiny pit containing a small number of black and white beads, each of which is about as wide as a human hair. The white beads are positively charged and the black beads negatively charged. Each pit is topped with a transparent electrode and has two other electrodes at its base. Altering the charge on the base electrodes makes either white or black beads leap to the top of the pit forming either a blank or black spot on the larger display. Making one base electrode positive and the other negative creates a grey spot.

Figure 2.15 Electronic paper is set to revolutionise display technologies.

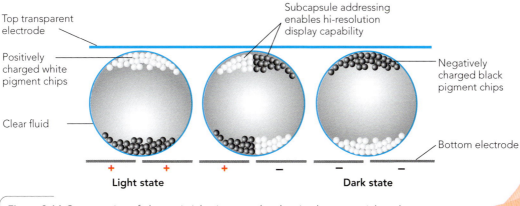

Top transparent electrode

Positively charged white pigment chips

Clear fluid

Subcapsule addressing enables hi-resolution display capability

Negatively charged black pigment chips

Bottom electrode

Light state

Dark state

Figure 2.16 *Cross-section of electronic-ink microcapsules showing how smart ink works.*

Radio frequency identification

Radio frequency identification (RFID) is a method of identification that uses tags stuck on to a product to store data that can be retrieved by a reader. This method is often used to track pallets of products from the manufacturer to the retailer. It also has applications in libraries, where a tag is used on library books to identify the book and lender and proves useful as a security system preventing books from being stolen.

RFID technology is based on the transmission and reception of radio frequency (RF) signals between a transmitter (known as a *reader*) and a transponder (known as a *tag*), which is attached to the product. In most cases the transmission is two-way: the transmitter sends signals, which the transponder receives; the transponder then transmits a response signal that is received by the transmitter. The information from the transmitter can then be used to identify the transponder and any item it is attached too.

RFID readers and tags work in many frequencies and are available in either active or passive formats.

- **Active transponders** are battery powered and can be read over a greater range, in some circumstances several tens of metres. However, they are typically expensive and as such are generally used on items such as vehicles (road toll payments) and shipping containers. Being battery powered they have a limited lifetime.

- **Passive transponders** take the power they to need to respond from the electromagnetic signals transmitted by the reader. However, the strength of these signals falls off quickly, giving limited read ranges. They are commonly used in applications where the transponder needs to be thin and easily converted into a label, such as pallet tracking or security systems in libraries.

Figure 2.17 *A radio frequency identification (RFID) tag and reader.*

THINK ABOUT THIS!

There are a growing number of people who believe RFID tags could be used for the wrong reasons. If RFID tags are attached to all products then they can be easily tracked – even back to your own home. What effect does this have on your privacy? How could this information be used by manufacturers and marketing companies?

2 Components

Binding methods

Binding is a process used to fasten or hold together a number of printed sheets. Many products are bound, such as magazines, books, reports and brochures. Binding can range from the simplest forms, for example stapling or ring binding, to fully automated processes. There are various methods of binding to choose from depending upon the specific application. For example, aesthetic considerations, the quantity of paper to be bound and the cost are determining factors as to which process is used.

Table 2.9 Binding methods for paper and board.

Method	Diagram	Applications	Advantages	Disadvantages
Saddle-wire stitching – the simplest method of binding, by stapling the pages through the fold.		Brochures, weekly magazines, comics.	• Ideal for signature feed processes (folded pages). • Printed materials can be laid flat to read. • Relatively inexpensive when produced commercially.	• Lower-quality visual appearance. • Not durable as centre pages can easily fall apart.
Side-wire stitching – staples are passed through the side of the document close to the spine.		Many modern photocopiers can collate and staple documents, e.g. information booklets, revision materials, etc.	• Used when the document is too thick for saddle-wire stitching. • Relatively inexpensive when produced commercially. • Ideal for binding multiples of single sheets of paper without folds.	• Cannot lay printed materials flat to read as it causes damage to spine. • Lower-quality visual appearance.
Spiral and comb binding – pages are punched through with a series of holes along the spine. A spiralling steel or plastic band is inserted through the holes to hold the sheets together.		Business reports/documents.	• Relatively inexpensive when produced commercially. • Ideal for binding multiples of single sheets of paper without folds. • Fairly good-quality visual appearance. • Printed materials can be laid flat to read.	• Not durable as document can easily fall apart or tear down perforations.
Perfect binding – pages are held together and fixed by the cover by means of a flexible adhesive.		Paperback books, glossy monthly magazines, catalogues.	• Better presentation and visual appeal with printable spine rather than staples. • Better quality – puts all the pages or signatures together, roughens and flattens the edge, then a flexible adhesive attaches the paper cover to the spine. • Glued spine provides longevity for a monthly magazine.	• Expensive commercial process.
Hard-bound or case-bound – usually combines sewing and gluing to create the most durable method of commercial binding.		Hardback books, quality presentations, e.g. school yearbooks.	• Stiff board is used on the cover to protect the pages. • High quality, professional binding method. • Extremely durable.	• Very expensive commercial process.

Industrial and commercial practice

Getting started!

As a designer, you need to know how products are manufactured. It's quite simple – you need to be able to design products that can be made and to make products that you have designed. What would happen if designers simply scribbled something on a piece of paper and gave it to somebody else to make? Would the manufacturer understand the drawing and, more importantly, what could go drastically wrong if information is not communicated effectively?

1 Scale of production

The scale of production is an important factor to be considered when developing any product. It has an impact upon all design and manufacturing decisions, including:

- the number of products or units manufactured

- the choice of materials and components

- the manufacturing processes, speed of production and availability of machinery and labour

- production planning, the use of just-in-time (JIT) and stock control, including the use of information and communication technology (ICT) systems

- production costs, including the benefits of bulk buying, the use of standard components and eventual retail price.

One-off

One-off production is often referred to as job production and includes 'tailor-made' bespoke or customised solutions. A key feature of one-off production is a single, often high-cost product that is manufactured to a client's specification. This kind of one-off product is often very high cost because a premium has to be paid for any unique features, more expensive or exclusive materials and time-consuming handcrafted production and finishing.

WEBLINKS: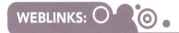

www.nikeid.com – Customised sports products

Figure 2.18 Multinational companies such as Nike have seen the benefit of providing customised products as well as their usual batch and mass-produced products.

Batch

Batch production involves the manufacture of identical products in specified, pre-determined 'batches', which can vary from tens to thousands. A key feature of batch production is flexibility of tooling, machinery and workforce to enable fast turnaround, so production can be quickly adapted to manufacture a different product, depending on demand. Batch production often makes use of flexible manufacturing systems (FMS) to enable companies to be competitive and efficient. The use of computer integrated manufacturing (CIM) systems involving automated machinery enables production 'downtime' to be kept to a minimum. Batch production results in lower unit cost than one-off production. Economies of scale in materials buying enables cost savings and identical batches of consistently high-quality products are manufactured at a competitive cost.

Figure 2.19 Mass production incorporating specialised tasks is a feature of the electronics industry.

LINKS TO:

Unit 2: Quality: Quality control in print runs

Table 2.10 Batch production in short-term print runs e.g. full-colour brochure.

Stage	Processes
Preparation: pre-press	• Colour separation of images into Cyan, Magenta, Yellow and Black (CMYK). • Production of four printer negatives for making printing plates. • Plate-making machines produces four colour printing plates (CMYK). • Quality control of printing plates and colour proofs.
Processing/ production	• Full-colour printing process, e.g. offset lithographic printing. • Application of in-line surface finish, e.g. lamination, spot varnishing, etc. • Quality control using printers' marks, i.e. colour bars and registration.
Assembly	• Trimming pages to correct size on guillotine using crop marks. • Double pages folded down spine. • Pages collated and bound using saddle-wire stitching. • Final quality-control check of finished product.
Finishing	• Collation of multiple brochures, packing into corrugated board boxes and sealing. • Multiple boxes palletised and shrink-wrapped ready for distribution.

Mass

Mass production (or high-volume production) of most consumer products makes use of efficient automated manufacturing processes and a largely low-skilled workforce. However, this workforce is specialised and production is divided up in specific tasks with labour to match. Mass-produced products are designed to follow mass-market trends, so the product appeals to a wide national and international target market. Production planning and QC in production enables the manufacture of identical products. Production costs are kept as low as possible so the product will provide value for money.

Continuous

Continuous production is used to manufacture standardised mass-produced products that meet everyday mass-market demand. The production of a blow-moulded fizzy drinks bottle, for example, necessitates 24-hour production, 7 days a week to satisfy consumer demands for soft drinks. This type of production is highly automated and uses machines that can run continuously for long periods of time with breaks only for routine maintenance.

Table 2.11 *Applications, advantages and disadvantages of scales of production.*

Scale of production	Applications	Advantages	Disadvantages
One-off	Prototype and architectural models, shop signage, vinyl stickers for commercial vehicles, etc.	• Made to exact personal specifications. • High-quality materials used. • Highly skilled craftsperson ensures high-quality product.	• Expensive final product in comparison to larger scales of production. • Generally labour intensive and can be a relatively time-consuming process.
Batch	Commercially printed materials, e.g. magazines and newspapers.	• Flexibility in adapting production to another product. • Fast response to market trends. • Identical batches of products produced. • Efficient manufacturing systems can be employed. • Very good economies of scale in bulk buying of materials. • Lower unit costs.	• Poor production planning can result in large quantities of products having to be stored, incurring storage costs. • Frequent changes in production can cause costly re-tooling, reflected in retail price.
Mass	Electronic products, e.g. mobile phones and games consoles, commercial packaging, etc.	• Highly automated and efficient manufacturing processes. • Specialisation of workforce to specific tasks. • Rigorous quality control ensures identical goods. • Excellent economies of scale in bulk buying of materials. • Increased production means that set-up costs are quickly recovered. • Low unit costs. • Reduced labour costs.	• Low-skilled workforce – low wages, repetitive nature of tasks leading to job dissatisfaction. • Ethical concerns of manufacturing in developing countries i.e. 'sweat shops'. • High initial set-up costs due to very expensive machinery and tooling needs. • Inflexible – cannot respond quickly to market trends.
Continuous	Packaging, e.g. cans and bottles for the drinks industry.	• As mass production. • Extremely low unit costs. • Runs continuously 24 hours, 7 days a week.	• As mass production. • Very little flexibility at all as production set up 24/7.

2 Graphical communication

Graphical communication involves the visual representation of information. These graphics are used anywhere where information needs to be explained quickly or simply, such as in signage, newspapers and, now, prolifically on the Internet. In D&T, there are two common forms.

- **Pictorial drawings** consist of a number of methods of communication used to visualise and represent objects in 3D form, including exploded/assembly drawings.

- **Working drawings** are produced to communicate all of the required information of a product in 2D form so that it can be easily read, interpreted and manufactured by a third party.

Table 2.12 Pictorial and working drawing methods.

Drawing method	Example	Description	Application
Isometric		Isometric drawing is an accurate method of showing three faces of an object. • All lines are true or scaled, so foreshortening does not occur. • All vertical lines are drawn 90 degrees to the parallel. • All horizontal lines on the object are drawn at 30 degrees to the vertical using a 30/60-degree set square. • Circles appear slightly distorted, so are drawn as ellipses.	Ideal for sketching design ideas and illustrating product designs.
2-point perspective		2-point perspective allows an object to be drawn as the human eye views it. Parallel lines appear to converge at a vanishing point (VP) the further away they are. • Two vanishing points are used on a horizon line: one for length and one for width. • Lengths, heights and widths will appear to foreshorten as they recede into the distance, using either the "division" or "addition" method to make the foreshortening more accurate. Where the object is drawn in relation to the horizon will determine what view is required. • **Worm's eye view** – the object is above the horizon so the underneath can be seen. • **Street level** – the horizon line passes through the object giving a dramatic effect. The top and underneath cannot be seen. • **Bird's eye view** – the object is below the horizon line allowing the top to be seen.	Widely used method for illustrating product designs, architecture and interior designs.
Planometric (axonometric)		Planometric, also known as axonometric, technique is often used for interiors as it allows a bird's eye view of the object. • The plan is drawn 45/45 degrees or 30/60 degrees to the horizontal. • All dimensions are true or in scale. Vertical lines can be reduced by three-quarters to one-third to avoid any distortion. • Circles and curves are drawn as they appear in view i.e. a circle is drawn with a compass or circle template.	Best suited to architectural drawings, especially interior design.

Drawing method	Example	Description	Application
3rd angle orthographic projection		3rd angle orthographic projection is a means of representing a 3D object in 2D. It uses multiple views of the object, including: • **front elevation** – the view looking at the front of the object • **end elevation** – which side the end is viewed from determines whether the end elevation is drawn to the left or to the right of the front elevation • **plan** – the view from above (bird's eye view) • **hidden detail** – details such as holes, slots and grooves that cannot be seen from the particular view • **section views** – taking a slice through an object and drawing what you see so all internal details can be shown. 3rd angle orthographic projection should be drawn to British Standards Institute (BSI) standards including: • **dimensioning** – the dimension should be written in milimetres and should be written on top of the dimension line • **centre lines** – broken lines should be drawn through the centre of circles and down the length • **3rd angle symbol** – this should be visible on the drawing to signify to any third party that it is a 3rd angle orthographic projection (as opposed to 1st angle orthographic).	Engineering drawings, planning for models and prototypes.
Nets (developments)	For an example of a net, see Fig 2.23 overleaf.	A net, also known as a development, is a flat 2D shape than can be cut, scored and folded to produce a 3D shape. A net must include: • **cut lines** – a continuous line where the material is to be cut • **fold lines** – a broken line where the material is to be scored, folded, bent or heated. • **tabs** – essential constructional information especially for paper, card and board showing where glue is to be applied or where dust flaps or tucks are required. • **a closure system**, such as a tuck flap or crash base.	Structural packaging design.

LINKS TO:

Unit 2: Industrial and commercial processes:
Designing and creating packaging nets

Unit 1: Portfolio of creative skills:
Product Design: Communication

Translations

Translation, in graphical terms, is the interpretation of one method of drawing of an object into another.

- **Translation from working drawings to pictorial drawings**, e.g. studying the 3rd angle orthographic projection and drawing a 2-point perspective to represent the same object.

- **Translation from pictorial drawings to working drawings**, e.g. isometric to 3rd angle orthographic projection (Figures 2.20 and 2.21).

- **Translation from pictorial drawings to nets**, e.g. studying an isometric drawing of a piece of packaging and drawing what the net would look like in order to construct the package (Figures 2.22 and 2.23).

- **Translation of nets to pictorial drawings**, e.g. 2D net to 3D isometric.

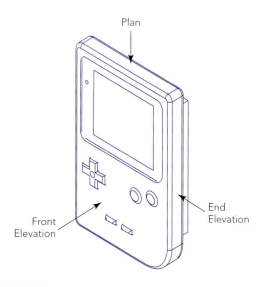

Figure 2.20 Pictorial (isometric) drawing of a hand-held games console.

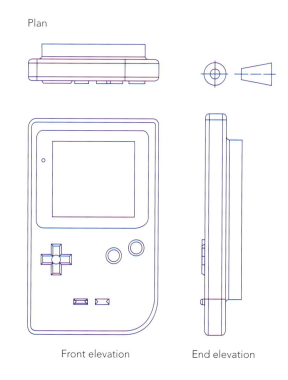

Figure 2.21 Third angle orthographic translation of pictorial drawing of hand-held games console.

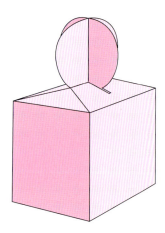

Figure 2.22 Pictorial drawing of a piece of packaging.

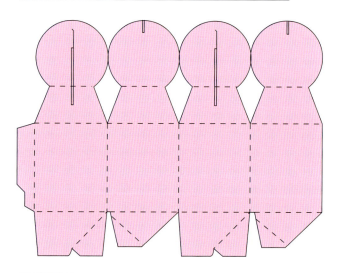

Figure 2.23 Translation of packaging into a net (development).

It is important that you practice all of these drawing methods through a series of drawing exercises using drawing boards and equipment. These drawing methods can be used in your portfolio of creative skills. However, in the exam you will not have a drawing board so you must also practice freehand sketching of these methods.

3 Computer-generated graphics

The use of computers for designing printed materials and products is now the industry standard, replacing the need for laborious and time-consuming manual drawing and cut and paste techniques. The possibilities created by this new technology have led to rapid developments in printed media, the Internet, animation and special effects for TV and film. Central to this new world of computer-generated graphics has been the development of Internet web sites and computer games which are now a key part of the graphic design industry. As with the use of all ICT-based systems, designers have had to re-train and develop new design skills. The extremely steep learning curve required when learning a complicated new software package is one of the main drawbacks of these systems.

Desktop publishing

Desktop publishing (DTP) combines the features of word processing, graphic design and printing all in one package. The modern print media all use DTP to produce the layout of their pages digitally rather than the traditional process of typesetting. Here are some features of the simplified DTP process for producing a page layout.

- Text can be imported from a word-processing package or typed directly into the DTP package.

- Illustrations or graphics can be created within a specialised package, uploaded from a camera or scanned and manipulated as required.

- The text and graphics are combined within the DTP package and inserted into a predetermined grid layout. Elements can be cropped or manipulated to best fit the page layout and add visual impact.

- A proof copy of the page layout can be printed in colour for evaluation and/or modification.

Advantages of DTP for modifying designs:
- text and image manipulation
- wide availability of typefaces
- graphic tools available
- page layout grid and guides can be easily added on screen (invisible when printed)
- cut and paste facility to alter page layouts
- ability to zoom in and out for attention to detail.

Disadvantages of DTP for modifying designs:

There are very few disadvantages which specifically relate to DTP software other than the general disadvantages of using computers. These are:

- the need to regularly invest in new software and hardware and the consequent re-training needed
- incompatibility between different software file types and computer systems.

2D computer graphics

2D computer graphics and digital illustration are rapidly growing areas in the field of design. Many drawings that would have traditionally been hand drawn and rendered can now be easily achieved with the use of the appropriate software. These 2D software packages offer a range of tools to construct and manipulate images. The basic tools allow shapes and lines to be drawn, to which colour and texture can then be added. Effects and filters can be applied to vary the look such as blur or mosaic. Digitised images, such as scanned pictures and existing graphics files, can be combined and manipulated in the same way.

Many of the contemporary characters designed for T-shirts, album covers, illustrations etc., are simply constructed using basic geometric shapes layered on top of each other with different colour fills.

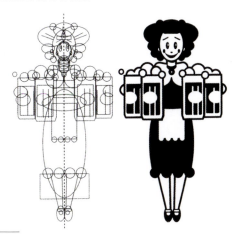

Figure 2.24 *Creating a 2D computer graphic from simple geometric shapes.*

THINK ABOUT THIS!

Sometimes we take computers for granted – how did we ever manage without them! Take a word-processing package like Microsoft Word for instance. What features does it have that make life easier for you when producing essays or reports? How does this compare with the traditional drafting and re-drafting of work on paper?

4 Modelling and prototyping

Sketch modelling

Sketch modelling helps to determine shape, dimensions and surface details by constructing a 3D representation of the concept product, usually out of foam or a similar material. Sketch models can be extremely useful in determining the ergonomic factors of many products. By constructing a number of models of varying shapes and sizes it is possible for designers to literally get a 'feel' for the product. It will soon become apparent in 3D form which designs are aesthetically pleasing or 'user friendly' and are worth developing – something that 2D images struggle to achieve.

Rapid prototyping using CAD/CAM

The need for manufacturing industries to cut down on the time and costs involved in developing a new product has led to the development of rapid prototyping (RPT). This involves the creation of 3D objects using laser technology to solidify liquid plastic polymers or resins in a process called stereolithography. Using specialist software applications that can be downloaded on to a stereolithography machine, 2D CAD drawings are converted to 3D models. The process is based on the computer slicing the 3D object into hundreds of very thin layers (typically 0.125–0.75mm thick) and transferring the data from each layer to the laser.

The laser draws the first layer of the shape on to the surface of the resin, which causes it to solidify. The layer is supported on a platform that moves down enabling the next layer to be drawn. This process of drawing, solidifying and moving down quickly builds up one layer on top of another until the final 3D object is achieved. Most companies do not have this technology. Instead, they use RPT services from specialist companies. Stereolithography prototypes can typically be delivered within 3–5 days of the client's design data being received, therefore saving both time and development costs.

Figure 2.25 *Block modelling at the early stage of the development of the Dyson vacuum cleaner.*

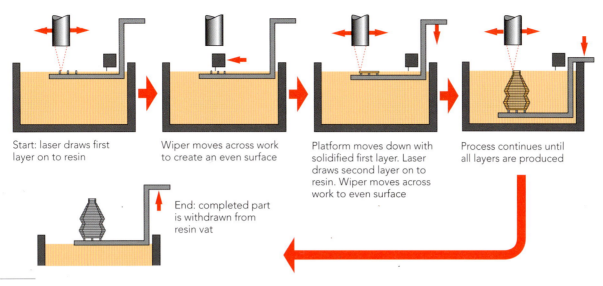

Start: laser draws first layer on to resin

Wiper moves across work to create an even surface

Platform moves down with solidified first layer. Laser draws second layer on to resin. Wiper moves across work to even surface

Process continues until all layers are produced

End: completed part is withdrawn from resin vat

Figure 2.26 The process of stereolithography.

The main disadvantage of RPT with stereolithography is the cost. A typical stereolithography printer will cost in excess of £10,000. Although the cost of each model is determined by factors such as its size and complexity, it will likely run into hundreds of pounds.

Another RPT system is selective laser sintering (SLS), which uses a high-power laser to fuse small particles of polymer (usually nylon), metal, or ceramic powders into a mass representing a desired 3D object. The laser selectively fuses powdered material by scanning cross-sections generated from a 3D digital description of the part on the surface of a powder bed. After each cross-section is scanned, the powder bed is lowered by one layer thickness, a new layer of material is applied on top, and the process is repeated until the part is completed.

Advantages of using RPT in the development of products.

- Fast entry to marketplace due to reduced lead-time between design concept and the reality of components reaching market or approval.
- Produces very complex and intricate shapes directly from CAD data.
- Reduces development time and therefore saves money.
- Accurate prototyping, as computer design data is replicated in detail, so complex shapes are easily produced.
- Accurate testing, as materials in the model are more representative of the final product.

Virtual modelling

A 3D image can give a more realistic impression than a 2D image. Therefore, many designers will construct new products on-screen that, with skill, can easily be modified and manipulated. Product design teams can significantly decrease the time taken to design and develop a new product with 3D modelling, saving development costs and reducing time to market. These virtual products can be tested and evaluated without actually being manufactured and design data can be directly output to a CAM or RPT system for modelling prototypes.

The rapid development in computer gaming has produced an impressive range of computer-generated characters interacting with life-like scenarios and landscapes. Rarely is a feature film released nowadays without containing computer-generated special effects and even actors. 3D animation starts with a wire frame model being created on-screen that can be viewed in all directions using built-in camera angles in the computer software. The wire frame model can then be rendered, giving it an appropriate surface finish or texture. This involves wrapping a 'skin' around the wire frame to give a photo-realistic image. Photo-realistic virtual models are often used by architects to show clients what their new building will look like. Animated 'walk-throughs' of interior designs give the client a real feel for the finished product.

THINK ABOUT THIS!

Discuss the advantages of using virtual 'walk-through' tours by architects when showing a client a new building or interior. Remember, the client is usually not a fully trained architect or designer and needs information to be communicated clearly.

Figure 2.27 *Architects made extensive use of virtual modelling when planning the London 2012 Olympic park.*

Table 2.13 *3D modelling techniques.*

Modelling technique	Example	Advantages	Disadvantages
Wire frame model – a wire frame model is a visual representation of a 3D object created using a collection of straight lines, curves and intersections. The result is a transparent 'skeleton' of a drawing that gives an impression of the overall 3D structure.		• Relatively simple and fast to calculate on a computer. • Used in cases where a high screen frame-rate is needed, e.g. planning CGI animation. • Able to review changes or rotate the object without long delays. • Well suited and widely used in computer numerically controlled (CNC) programming.	• Difficult to read or interpret as all hidden lines are also in view. • Lacking in detail, i.e. no surface finishes.
Surface model – a surface or 'skin' is added to the wire frame model, making it a more realistic picture of the 3D object.		• Greater definition of contours and surfaces. • Blending between surfaces is improved, i.e. how smoothly they connect to one another. • Hidden lines can be removed to give a more solid feel.	• Does not give any information about surface finishes or textures.
Solid model – the production of a full digital representation of an object complete with the properties of the solid form.		• Photorealistic image produced using rendering techniques to apply shading, surface pattern and texture. • Clear communication – no confusion.	• Complex data required, therefore slow to update and render objects.

5 Joining techniques

Preparation

Adhesives require clean and thoroughly prepared surfaces in order for the joint to fully bond. As polymers often have a smooth and shiny surface finish, they should firstly be cleaned and roughened with an abrasive paper before bonding. Metal surfaces should also be degreased and roughened with an abrasive paper. Woods are porous so the adhesive will 'soak' into its surface. Once the adhesive has been applied to the cleaned and prepared surface it then has to cure (harden). It is important that the two surfaces are held in place securely so that they do not slip as this process can take several minutes, or hours.

LINKS TO: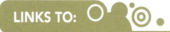

Unit 2: Health and Safety: Control of Substances Hazardous to Health (COSHH) regulations.

Table 2.14 *Adhesives for joining like and unlike materials.*

Adhesive	Applications	Advantages	Disadvantages
Contact adhesive	Metals and polymers; unlike materials, e.g. plastic to wood; general purpose; fabric to most materials	• Able to bond unlike materials. • Bond forms very quickly so less need for clamping, etc. • Ideal for gluing large sheet materials. • Sold in metal tubes for easy application.	• Must be applied to both surfaces and some time must be allowed before they can be pushed together. • Relatively expensive. • Solvent-based so contains harmful volatile organic compounds (VOCs).
Acrylic cement	Acrylic	• Rapid bonding of acrylics.	• Can be difficult to apply and give a neat joint. • Relatively expensive. • Solvent-based so contains harmful volatile organic compounds (VOCs).
Polystyrene cement	High-impact polystyrene (not expanded polystyrene)	• Strong bond – melts surface of pieces to be joined and causes them to weld together. • Able to use a brush to apply (water-like consistency) and absorbed into joint by capillary action.	• Relatively expensive. • Solvent-based so contains harmful volatile organic compounds (VOCs).
Epoxy resin	Most materials including expanded polystyrene	• High-performance adhesive giving high-strength bonds. • Chemical reaction hardens immediately. • Versatile – can be made flexible or rigid, transparent or opaque/coloured, rapid or slow setting. • Excellent heat and chemical resistance.	• Reaches full strength only after a few days. • Expensive. • Often requires manual mixing of resin and hardener, which can be messy.
Polyvinyl acetate (PVA)	Woods (and porous materials)	• Gives a strong joint. • Relatively inexpensive.	• Surfaces need to be securely clamped together for long periods in order for PVA to harden. • Generally not waterproof (although some brands are).

6 Industrial and commercial processes

Structural packaging nets

A net, also known as a development, is a flat 2D shape than can be cut, scored and folded to produce a 3D shape. Nets are widespread in the production of packaging using cartonboards. The packaging of a product is extremely important as it serves four main functions.

- **To contain** the product safely while in transit using suitable materials and containers.

- **To protect** the product while in transit to avoid breakages and unnecessary wastage by using suitable materials.

- **To dispense** the product in a safe and convenient manner by using suitable closures.

- **To advertise** the product for retail purposes in order to initially attract the customer's attention and then provide essential information.

Many companies produce standard packaging nets for designers to adopt and simply add individual graphic identity. This ultimately will save valuable time and costs in developing a product to market. There is also an internationally recognised system using diagrams of net constructions and symbols that avoids the need for lengthy and complicated verbal descriptions, especially useful in the global marketplace.

Designing and creating packaging nets

The accuracy of the drawing is extremely important when constructing nets. When drawing nets, technical drawing equipment or CAD programs should offer accuracy and consistency. To produce an accurate net the final 3D shape will need to be developed and drawn either by hand or on computer. This will enable the shape, size and layout of the net to be drawn more easily. The net will need to show the following constructional information:

- cut lines – a continuous line where the material is to be cut

- fold lines – a broken line where the material is to be scored, folded, bent or heated

- tabs – essential constructional information especially for paper, card and board showing where glue is to be applied or where dust flaps or tucks are required

- the closure system

- annotation – will assist in labelling edges, sides and features in relation to one another.

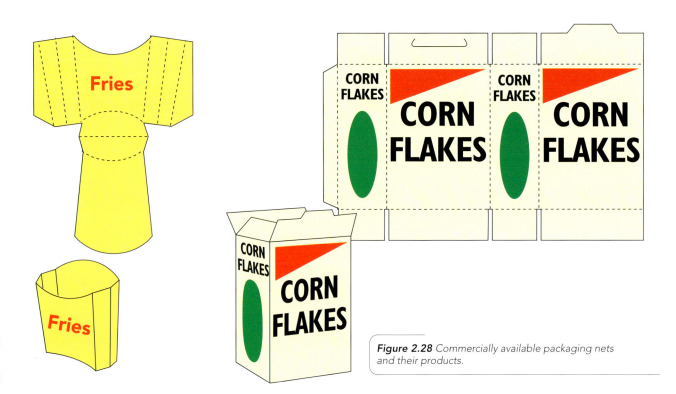

Figure 2.28 *Commercially available packaging nets and their products.*

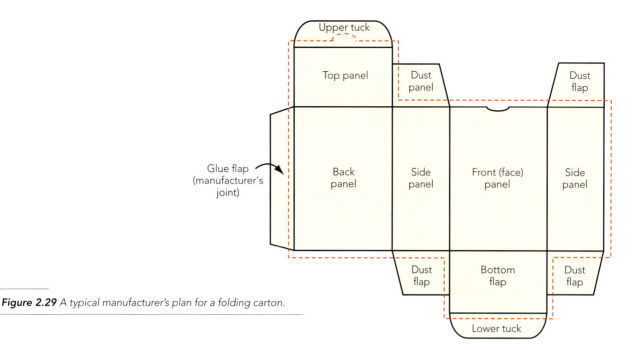

Figure 2.29 *A typical manufacturer's plan for a folding carton.*

THINK ABOUT THIS!

After you have finished with a carton, don't simply throw it away – have a go at quickly sketching what you think the net looks like. The answer is easily viewed by simply undoing the carton to reveal the net.

Commercial production of packaging nets

The commercial production of packaging nets often requires large-scale production processes involving the use of automated machinery.

Making the die form

Packaging nets for cartons need to be cut to shape after printing and before assembly or gluing. The cutters used are called dies and the stamping process is called die-cutting. Making a die-form can be carried out using CAM-operated forming machines but is often a skilled, hand-crafted process. The process involves cutting and shaping hardened steel rules (or press knives) and matrix strips that will either cut or crease the cartonboard on the die-cutting machine. Ejector rubbers are added to the cutting rulers to push the cartonboard away from the rulers after a cut has been made. The die-form is then mounted on the die-cutting machine. The making of the die-form is the major cost in the setting up of the carton-making job.

Table 2.15 *Stages in the commercial production of packaging nets.*

Stage	Processes
Preparation	• Making die form for die cutting and creasing process. • Pre-press – making four colour printing-plates (CMYK). • Purchase of cartonboard, printing inks and adhesives.
Processing/production	• Full colour printing onto sheets of cartonboard, i.e. offset lithographic printing. • Application of in-line surface finish, e.g. lamination, spot varnishing, etc. • Die-cutting of individual nets incorporating scoring of fold lines.
Assembly	• Folding of net into shape. • Gluing of tabs to form 3D structure. • Product and internal packaging inserted and then final flap glued into position.
Finishing	• Collation of multiple units, packing into corrugated board boxes and sealing. • Multiple boxes palletised and shrink-wrapped ready for distribution.

Figure 2.30 An example of a die-cut box in a flat state.

Die cutting

While the die-form is being made the cartonboard sheets for the job are selected and cut to size on a power guillotine. The cartonboard is then printed using the appropriate printing process, such as offset lithography.

After the die-form is mounted on the machine, a test sheet is cut and adjustments and corrections made. Once satisfied, each printed sheet is fed onto a platen, directly below the die-form. The platen is then forced upwards, pushing the cartonboard hard against the rulers. The cutting rules cut right through the cartonboard, while the creasing rules push the card into the groove formed by the matrix strips. Once cut, the sheets are pulled to the next stage to be stripped of their waste. Ejector pins on the die-form push the waste away onto a conveyor belt for disposal while the nets leave the machine and are stacked ready for gluing.

Folding and gluing

In mass production or large batch production the nets are folded and glued on a highly automated gluing line. The first operation is to pre-fold one of the creases, and then the carton will travel to an automatic gluing station. The gluing module is 'timed' by a simple control system. When the front edge of the card blank breaks a light beam it triggers a device that squirts tiny drops of glue in quick succession. The final fold is made and the edges are held together under a moving belt until they reach the final stage. The cartons are stacked up and compressed, using

carefully controlled pressure, under another pair of slow-moving rubber belts. Here the adhesive has time to cure before the cartons are removed and packed for transit. The cartons are then shipped to the customer's site for assembly and filling.

THINK ABOUT THIS!

It is a fascinating process to view packaging being produced by machinery. Try to locate a suitable company in your local area and organise a trip to their premises. You will be amazed at what modern automated machinery is capable of.

7 Forming techniques

Blow moulding

In the blow moulding process a hollow thermoplastic tube (the parison) is extruded between a split mould and clamped at both ends. Hot air is blown in to the

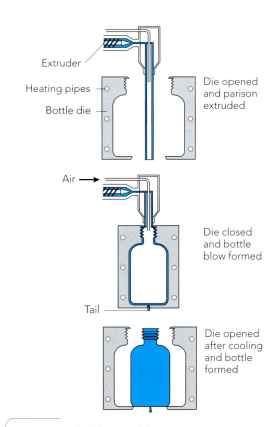

Figure 2.31 The blow moulding process.

parison, which expands to take the shape of the mould, including any relief details such as threads and surface decoration. Once the polymer cools and solidifies the product is ejected by opening the split mould. Blow moulded containers do not have to be symmetrical and can incorporate handles, screw threads and undercut features.

Injection moulding

The injection moulding process makes use of a high-cost mould that is injected with a liquid polymer made by heating thermoplastic granules. Once the polymer cools and solidifies the formed product is ejected. Injection moulding is suitable for complex shapes with holes, screw fittings and integral hinges, formed by thinning the polymer.

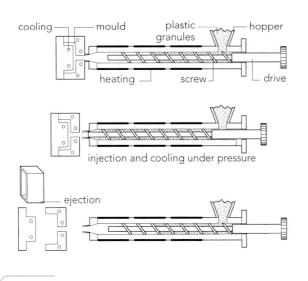

Figure 2.32 The injection moulding process.

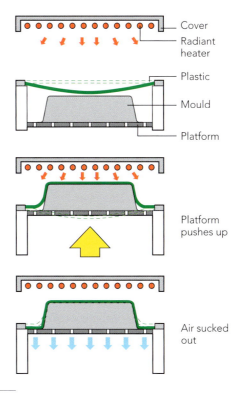

Figure 2.33 The vacuum forming process.

Vacuum forming

In vacuum forming, a thermoplastic sheet is clamped and heated, blown and stretched. Air is sucked out of the vacuum forming machine to pull the softened sheet over a mould pushed up from below. Once the polymer has cooled it solidifies and cold air is blown between the mould and thermoplastic sheet to release the formed product. In industry, 'multi moulds' of four, six or 12 identical moulds are used to create small batches in one forming.

Table 2.16 Advantages, disadvantages and applications of thermoforming techniques.

Process	Advantages	Disadvantages	Polymers used	Applications
Blow moulding	Intricate shapes can be formed. Can produce hollow shapes with thin walls to reduce weight and material costs. Ideal for mass production – low unit cost for each moulding.	High initial set-up costs as mould expensive to develop and produce.	HDPE, LDPE, PET, PP, PS, PVC	Plastic bottles and containers of all sizes and shapes, e.g. fizzy drinks bottles and shampoo bottles.
Injection moulding	Ideal for mass production – low unit cost for each moulding for high volumes. Precision moulding – high-quality surface finish or texture can be added to the mould.	High initial set-up costs as mould expensive to develop and produce.	Nylon, ABS, PS, HDPE, PP	Casings for electronic products, containers for storage and packaging.
Vacuum forming	Ideal for batch production – inexpensive. Relatively easy to make moulds that can be modified.	Mould needs to be accurate to prevent webbing from occurring. Large amounts of waste material produced.	Acrylic, HIPS, PVC	Chocolate box trays, yoghurt pots, blister packs, etc.

8 Finishing processes

Enhancing the format of paper and board

There are a number of finishing processes that can be used to improve the performance and quality and enhance the aesthetic and functional properties of paper and board. However, all of these processes place additional costs onto a print job.

Laminating

Laminating applies a transparent plastic film to the surface of paper and board. Commercial laminating uses a polypropylene (PP) film that is glued to the paper as it is fed through a heating wedge under high pressure. Lamination provides a wide range of uses across the whole spectrum of printed products due to its properties of good gloss and strength and the advantages of low cost.

Encapsulation

Encapsulation is very similar to laminating with the addition of heat seams, therefore fully covering the edges of the paper or board. Commercial printers will use roll laminators for large print runs such as menus for restaurant chains. They require menus to be encapsulated to prevent them from becoming creased or wrinkled and easily wiped clean from food stains, or marked by grease from fingerprints.

When smaller batches are required, for example, identity (ID) cards for schools or libraries, small and inexpensive pouch laminators are used. The inside of the lamination pouch is coated with a heat-activated film that sticks to the paper as it runs through the laminator. Inside the laminator the paper passes through a heater to activate the adhesive and then through rollers under pressure, ensuring that all adhesive layers bond to one another and that the edge is sealed.

Varnishing

Varnish is applied to paper and board in order to give it a high-gloss finish, for example, the pages of glossy magazines to give the paper a quality feel. The process involves a fine varnish being sprayed on to the surface of the paper or board. Once dry this gives a gloss finish that helps to protect the printing underneath. The varnishing process can only take place after the colour printing is completed. This is due to the oil- or water-based varnishes that are used, which take at least two hours to dry. This is a major disadvantage, as other finishing processes have to be delayed as the varnish dries. For example, a brochure cannot be collated, folded and bound as the pages would stick to one another.

To speed up the drying process, ultra-violet (UV) varnishing can be used. Special varnishes dry almost straight away if they are exposed to UV light. The varnish is sprayed onto the paper in the same way as other varnishes. However, after spraying the paper passes underneath UV lights, which dry the varnish almost instantaneously. This allows printed materials to move quickly on to other finishing processes. One disadvantage is that this type of machinery is expensive to purchase. However, it produces the ultimate in gloss finishes to paper and board. Varnish can also be applied using the screen printing process – and is particularly applicable to spot varnishing.

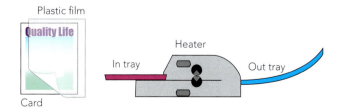

Figure 2.34 *Encapsulation using a pouch laminator.*

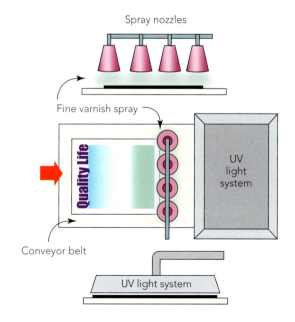

Figure 2.35 *The ultra-violet (UV) varnishing process.*

UV varnish can be applied on spot locations of the paper as well as flooding the whole page. Spot varnishing applies UV varnish to selected areas of a printed image to enhance product impact or form part of the graphic design. There are several types of UV varnishes that can provide even greater visual impact and functional properties.

- **UV sparkle varnish** containing metallic polyester flakes that adds "sparkle" when applied to selected areas of a printed image and will provide "shelf appeal" to a wide range of printed products.

- **Fragrance burst inks** are designed for 'scratch 'n' sniff' applications to selected areas of printed sheets. The fragrance is released by gently rubbing the fragrance burst area.

- **Silver latex "scratch offs"** are designed as opaque inks that can be printed over printed tickets or promotional cards and are readily removable using a coin or fingernail to display a pre-printed message underneath.

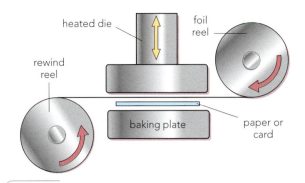

Figure 2.37 *The hot foil blocking process.*

Hot foil blocking

Hot foil blocking, also known as foil blocking or hot foil printing, is used to produce true 'reflective metal' printing and other effects impossible with normal metallic printing inks. As a result, hot foil blocking can be used to enhance and add value to conventionally printed materials. The hot foil blocking process transfers a foil coating to paper or board by means of a heated die. A roll of foil with a polyester backing sheet is continuously fed over the paper or board and a heated die presses the foil on to its surface.

Hot foil blocking can be used for marking a wide variety of surfaces, and has certain advantages over many other types of print finishes.

- Being a dry process, there are no problems with fumes or solvent vapours, or mixing inks, and the printed goods can be handled immediately.
- The printed image is 100% opaque.
- It can be very economic in short-run printing.

The main disadvantages of the process are:
- because the image is created under heat and pressure, it tends to 'spread' a little, which creates difficulties in reproducing tints and halftones; fine line work, however, can be printed satisfactorily
- the range of colours available, though wide, is limited to those produced by the foil manufacturers; while it may be possible to find a foil to match a custom colour requirement, in many cases the 'nearest available' will have to be used.

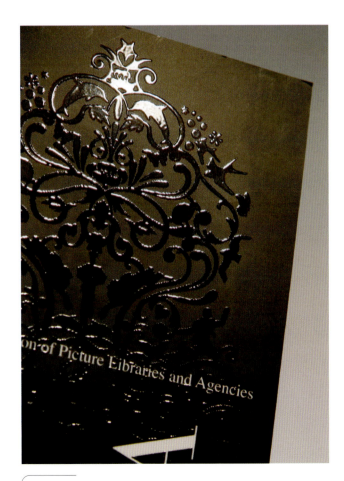

Figure 2.36 *UV spot varnishing to enhance product impact.*

Figure 2.38 *Hot foil blocking for visual impact.*

Embossing

Embossing is the process of creating a 3D image or design in paper and board producing a raised effect. Alternatively, using debossing, the image is lowered into the surface of the paper and board. The process involves the use of a heated metal die (female) and a counter die (male) that fit together under pressure and actually squeeze the fibres of the paper into the desired shape. This pressure and a combination of heat actually 'irons' while raising the level of the image higher than the paper to make it smooth. In commercial printing, embossing is carried out on a converted letterpress machine at the end of the production process after any varnishing or laminating processes. While the end result gives a high-quality and sophisticated appearance, the process costs as much as the printing, therefore doubling the cost of printed products. Confectionary products such as Toblerone and After Eight mints have embossed lettering on their packaging as do many greetings cards.

THINK ABOUT THIS!

Imagine you are promoting a new restaurant or nightclub and want to portray a highly professional and upmarket image. Evaluate the use of expensive printing effects for your promotional flyer. For example, is it worth the money in order to attract the right type of customer?

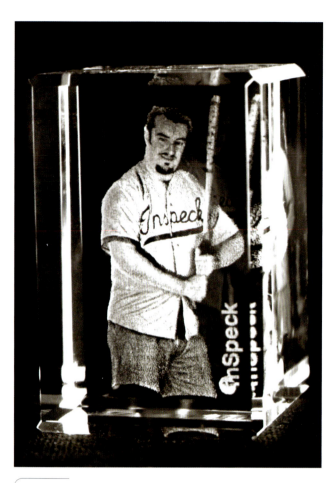

Figure 2.39 *Sub-surface laser engraving (SSLE) is the process of engraving an image below the surface of a solid material to produce 3D images in souvenir 'crystal' or promotional items.*

Surface decoration using CAD/CAM

Laser cutting and engraving

Laser cutting is a technology that uses a high-powered laser to cut materials, and is usually used in industrial manufacturing to produce features that otherwise would be problematic or even impossible to die-cut. It also eliminates the need for expensive tooling as the laser is directly controlled by computer using a CAD/CAM system. The laser either melts, burns or vaporises away the material leaving an edge with a high-quality surface finish. This process is ideal for manufacturing signage, for example, from sheet acrylic.

Laser cutters usually work much like a milling machine would for working in sheet materials. The laser (equivalent to the mill) enters from the side of the sheet and cuts it through the axis of the beam. In order to be able to start cutting from somewhere other than the edge, a pierce is made before every cut. Piercing usually involves a high-power pulsed laser beam that slowly makes a hole in the material.

Lasers are used in the printing industry in the following ways.

- **Direct laser imaging** of lithographic printing plates using a carbon dioxide laser imaging head to burn away unwanted material and form a negative printing plate directly from digital data.

- **Direct laser engraving** of flexographic and gravure printing plates and cylinders using a high-power carbon dioxide laser head to form sharp relief images with steep, smooth edges to give a high standard of process colour reproduction.

Advantages of laser cutting over mechanical cutting:

- lack of physical contact with the material produces a clean edge

- precision, as there is no wear on the laser and it is computer controlled

- reduced chance of warping the material that is being cut, as laser systems have a small heat affected zone

- some materials or features are very difficult or impossible to cut by more traditional means.

Disadvantages include:

- very expensive machinery

- high energy consumption required to power lasers.

Vinyl cutting

Vinyl is the common term used to describe plasticised PVC. Vinyl stickers or graphics are ideal for one-off or batch production, from an individual sign for a shop or restaurant frontage to a series of movie adverts covering the backs of buses. Many schools and colleges will have access to CAD/CAM facilities for cutting vinyl that directly replicate those of the commercial process. The image is designed on computer and digitally sent to the vinyl cutter for contour cutting. Commercial sign-makers usually have image banks and technically accurate dimensions for batch produced vinyl graphics, such as logos or trademarks for large companies.

Once the cutting is complete, all of the background vinyl is removed using a process called 'weeding', leaving only the required graphic. The adhesive used on rolls of vinyl is contact or impact adhesive that has been lightly coated on the back and covered with a treated backing paper to protect it and make the vinyl easy to peel off. Finally, a layer of application tape is applied over the graphic. The application tape will adhere to the vinyl so that it can be removed from the backing sheet and put onto the required surface. The nature of vinyl gives the letters a soft and flexible quality that allows them to be adhered to a variety of contours. Once applied, the application tape can be removed leaving only the vinyl graphic in place.

THINK ABOUT THIS!

Vinyl graphics are often used instead of the traditional hand-painted methods, especially on shop signage. Discuss the advantages of using CAD/CAM vinyl graphics for a shop sign. However, think about the impact upon the trade of the highly skilled professional sign writer.

Figure 2.40 Vinyl graphics have proved very successful when marketing and personalising the new Mini.

9 Printing processes

Commercial printing processes are distinguished by the method of image transfer used. Depending upon the process, the printed image is transferred to the paper either directly or indirectly.

- In **direct printing** the image is transferred directly from the plate cylinder (or image carrier) to the paper, e.g. gravure, flexography, screen printing and letterpress printing processes.

- In **indirect**, or **offset**, **printing**, the image is first transferred from the plate cylinder to the blanket cylinder and then to the paper, e.g. offset lithography, which is the most widely used commercial printing process.

Offset lithography

Lithography works on the basic principle that oil and water do not mix (they repel each other). Modern high-volume lithography is used to produce posters, books, newspapers, packaging, credit cards, decorated CDs – just about any flat surface, mass-produced item with print on it. The development of digital image setters has enabled print shops to produce negatives for plate making directly from digital images on computers using direct laser imaging – this is known as computer-to-plate (CTP). The positive image is the emulsion that remains after imaging.

The printing plate, made from a flexible aluminium or polymer, is fixed to the plate cylinder on a printing press. Rollers apply water, which covers the blank portions of the plate but is repelled by the emulsion of the image area. Ink, applied by other rollers, is repelled by the water and only adheres to the emulsion of the image area such as the text and photographs on a newspaper page. If this image was directly transferred to the paper, it would create a positive image, but the paper would become too wet. Instead, the plate rolls against a drum covered with a rubber blanket (blanket cylinder), which squeezes away the water and picks up the ink. The paper rolls across the blanket cylinder and the image is transferred to the paper. Because the image is first transferred, or offset to the blanket cylinder, this reproduction method is known as offset lithography or offset printing.

Offset lithographic presses involve multiple print units, each containing one printing plate for the four process colours of **C**yan, **M**agenta, **Y**ellow and Blac**k** (CMYK).

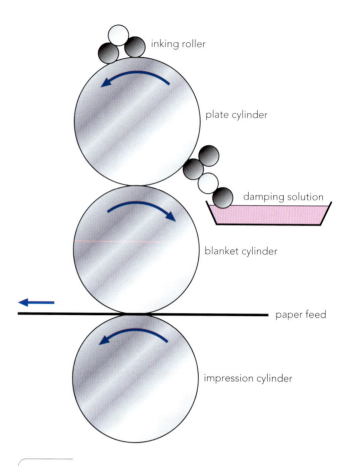

Figure 2.41 *The offset lithographic printing process.*

They are capable of printing multi-colour images in one pass on both sides of the sheet at very high speeds. Some presses can accommodate continuous rolls (webs) of paper, known as web presses. In order to produce richer colours, lithography presses often have 5 or even 6 separate colours. For example, when printing 'people' magazines, a pink or flesh tone is often added. There is no particular order with which the colours are printed. It is the printer who makes this decision depending upon the nature of the job and their understanding of how to produce the best finish.

A good way of seeing how the separate colours are used in lithography to create the range of colours is to place a page from a magazine under a microscope. You will see that the coloured areas are made up from thousands of tiny dots. This process is similar to pointillism within art, a technique made famous by George Seurat.

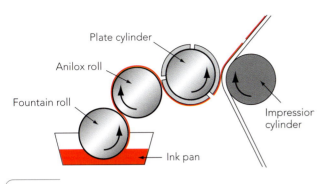

Figure 2.42 *The flexographic printing process.*

Flexography

Flexography is the main printing process used to commercially print packaging materials including cartonboard containers, plastic bags and chocolate bar wrappers. Flexography uses a relief-type printing plate with raised images and only the raised images come in contact with the paper during printing. Printing plates are made of a flexible material, such as plastic, rubber or UV-sensitive polymer (photopolymer), so that they can be attached the plate cylinder for ink application.

Flexographic presses comprise a plate cylinder, a metering cylinder known as the anilox roll that applies ink to the plate, and an ink pan. Some presses use a third roller as a fountain roller and, in some cases, a doctor blade for improved ink distribution. In the flexographic printing process, the paper is fed into the press from a roll. The image is printed as the paper is pulled through a series of print units. Each print unit prints a single process colour (CMYK). As with gravure and lithographic printing, the various tones and shading are achieved by overlaying the four basic shades of ink.

Screen printing

Screen printing is a widely used commercial printing process for producing many mass- or large-batch-produced graphics, such as posters or point-of-sale display stands. The screen is made of a piece of porous, finely woven fabric (originally silk, but nowadays typically made of polyester or nylon) stretched over a wooden or aluminium frame. A stencil is used to block off areas of the screen with a non-permeable material. This stencil is a negative of the image to be printed so the open spaces are where the ink will appear. Screens and stencils are produced commercially using the photo-emulsion technique.

- The original image is placed on a transparent overlay. The image may be drawn or painted directly on the overlay, photocopied, or printed with a laser printer, as long as the areas to be inked are opaque.

- The overlay is placed over the emulsion-coated screen and then exposed with a strong light.

- The areas that are not opaque in the overlay allow light to reach the emulsion, which hardens and sticks to the screen.

- The screen is washed off thoroughly. The areas of emulsion that were not exposed to light dissolve and wash away, leaving a negative stencil of the image attached to the screen.

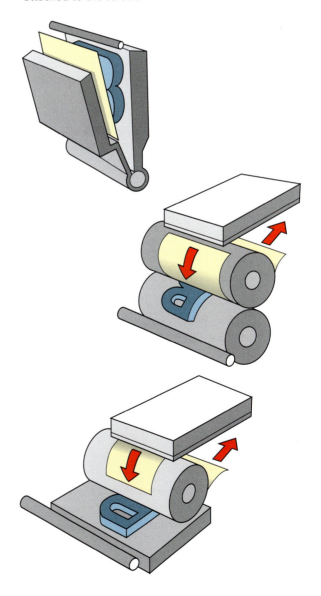

Figure 2.43 *The screen-printing process.*

The screen is placed on top of a piece of dry paper (or other material such as fabric). Ink is placed on top of the screen, and a squeegee (rubber blade) is used to push the ink evenly into the screen openings and onto the paper. The ink passes through the open spaces in the screen onto the paper below and then the screen is lifted away. The screen can be re-used after cleaning. If more than one colour is being printed on the same surface, the ink is allowed to dry and then the process is repeated with another screen carrying a different stencil and using a different colour of ink.

THINK ABOUT THIS!

Screen printing is one of the few commercial printing processes that can actually be replicated in school. Your art or textiles department may have some screens and printing inks for you to try this process for yourself. You could produce some T-shirts, which you might want to sell as a mini enterprise.

Gravure

Rotogravure (gravure for short) is a type of intaglio printing process where the image is engraved onto a copper plate cylinder. Gravure, like offset and flexography, uses a rotary printing press and the vast majority of presses print on reels of paper, rather than sheets of paper. Rotary gravure presses are the fastest and widest presses used commercially as they can print everything from narrow labels to 12-feet-wide rolls of vinyl flooring. In-line finishing operations such as saddle-wire stitching for magazines and brochures are also possible with a gravure press.

Plate cylinders can be engraved digitally by a diamond-tipped or laser etching machine. On the gravure plate cylinder, the engraved image comprises small recessed cells that act as tiny ink wells. The depth and size of these wells control the amount of ink that gets transferred to the paper.

A rotogravure printing press has one printing unit for each of the four process colours (CMYK). There are five basic components in each colour unit: an engraved plate cylinder, an ink fountain, a doctor blade, an impression cylinder, and a dryer. The plate cylinder is partially

immersed in the ink fountain, filling the recessed cells. As the cylinder rotates, it draws ink out of the fountain with it. Acting as a squeegee, the doctor blade scrapes the cylinder before it makes contact with the paper, removing ink from the non-printing (non-recessed) areas.

Next, the paper passes between the impression cylinder and the plate cylinder under pressure. Here, the ink is transferred from the recessed cells to the paper. The purpose of the impression cylinder is to apply force, pressing the paper onto the plate cylinder, ensuring even and maximum coverage of the ink. Finally, the paper passes through a dryer because it must be completely dry before going through the next colour unit and absorbing another coat of ink.

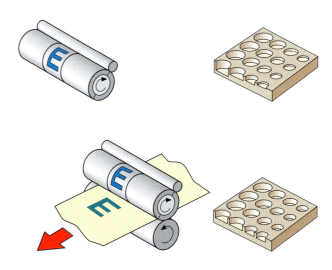

Figure 2.44 *The rotogravure printing process.*

Table 2.17 Advantages, disadvantages and applications of printing processes.

Process	Advantages	Disadvantages	Applications
Offset lithography	• Good reproduction quality, especially photographs. • Inexpensive printing process. • Able to print on a wide range of papers. • High printing speeds. • Widely available.	• Colour variation due to water/ink mixture. • Paper can stretch due to dampening. • Set-up costs make it uneconomic on short runs. • Can only be used on flat materials. • Requires a good-quality surface.	Business stationery, brochures, posters, magazines, newspapers.
Flexography	• High-speed printing process. • Fast-drying inks. • Can print on same presses as letterpress.	• Difficult to reproduce fine detail. • Colour may not be consistent. • Set-up costs high and would rarely be used on print runs below 500,000.	Packaging, less-expensive magazines, paperbacks, newspapers.
Screen printing	• Stencils easy to produce using photo-emulsion technique. • Versatile – can print on virtually any surface. • Economical for short, hand-produced runs. • Fully automatic methods capable of producing large volumes.	• Generally difficult to achieve fine detail (photographic screens able to reproduce fine detail). • Print requires long drying times.	T-shirts, posters, plastic and metal signage, point-of-sale displays, promotional items, e.g. pens, glasses and mugs.
Gravure	• Consistent colour reproduction. • High-speed printing process. • Widest printing presses. • Ink dries upon evaporation. • Variety of in-line finishing operations available. • Good results on lower-quality paper.	• High cost of engraved printing plates and cylinders. • Only efficient for long print runs. • Image printed as 'dots' that are visible to the naked eye. • Very expensive set-up costs, so only used on large print runs.	High-quality art and photographic books, postage stamps, packaging, expensive magazines.

Quality

Getting started!

What actually is quality? A friend might think that a certain product has quality but you might disagree, so, is it subjective? We can all name a product that we consider to be of a high quality but what qualities does it have? When designing products, quality refers to a product's ability to satisfy a need and, more importantly, its fitness for purpose. In manufacturing terms it is producing a product that is the best it can be, fully functional and free from defects.

1 Quality assurance systems and quality control in production

Quality assurance

Quality assurance (QA) systems are the planned activities used by the manufacturer to monitor the quality of a product from its design and development stage, through its manufacture, to its end use, and degree of customer satisfaction. In other words, QA is an assurance that the end product fulfils all of its requirements for quality.

- In the first instance, QA ensures a product is fit for purpose using thorough testing throughout the design and development stage.

- It includes regulation of the quality of raw materials and components that the manufacturer buys in order to start production.

- QA systems monitor the quality of components, products and assemblies in production through a series of quality control (QC) checks, tests and inspections processes.

- Finally, QA supplies fact-based evidence for quality management systems to inspire external confidence to customers and other stakeholders that a product meets all of their needs and expectations.

THINK ABOUT THIS!

We have all probably bought a product and found that it wasn't of a good quality. What do you suppose was the problem with it – design, materials, manufacture, assembly or finish? Was there a guarantee with it? Using your product analysis skills developed in Unit 1, outline the quality issues relating to a product that you are familiar with.

Quality control

QC is part of the achievement of quality assurance. It involves the actual activities used by a manufacturer to ensure a high-quality product is produced by means of inspection and testing.

Inspection

Inspection is the sampling and examination of components or products to check that they are within a specified tolerance. Tolerance is the degree to which a component is acceptable in order to function in accordance with its specification. For example, a drinks bottle must fit the machinery at the bottling plant for it to be held, moved and filled effectively, and must also hold the correct amount of liquid. A 54mm-diameter bottle is likely to have a tolerance of +/- 0.8mm. If, when inspected and tested, a bottle measures

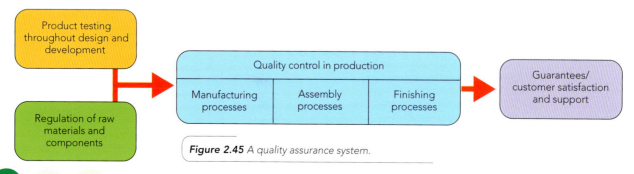

Figure 2.45 A quality assurance system.

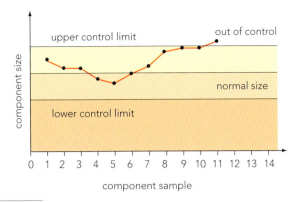

Figure 2.46 *Quality control chart using tolerance.*

Figure 2.47 *Testing paper tissue for tensile strength.*

between 53.2mm and 54.8mm, it would be within the agreed tolerance and would therefore be accepted. Any bottle that lies beyond this tolerance would be scrapped and recycled.

There are three main levels of inspection.

- **100% inspection** of all the units.
- **Normal inspection** using a sampling plan under ordinary circumstances. For example part of a sampling plan on a large print run of say 500,000 leaflets may involve every 1000th leaflet having its colours checked against the original to ensure consistency.

- **Reduced inspection** using a sampling plan requiring smaller sample sizes than those used in normal inspection. Reduced inspection is used in some inspection systems as an economy measure when the level of submitted quality is 'sufficiently good'. This may be used on a continuous print production line such as a company producing chocolate wrappers. Because the print runs 24/7, there is no need to check the print quality meticulously. In this case a visual check is sufficient.

Computer-aided inspection is possible by using a coordinate-measuring machine (CMM) for dimensional measuring. A CMM is a mechanical system designed to move a measuring probe to determine the coordinates of points on the surface of a workpiece. These machines are used to quickly and accurately measure the sizes and positions of features on mechanical parts, with tolerances as small as 0.0001 inches. Laser scanning systems are often used that can determine the coordinates of many thousands of points. This data can then be taken and used to not only check size and position, but to create a 3D model of the part as well using a CAD system. This technique is used for more complex 3D products and would not be used for the production of 2D graphic products where very close dimensional tolerances are rarely required.

Testing

Testing is concerned with the product's performance. Tests are carried out in laboratory conditions with strict control procedures to ensure that the results obtained are accurate. Tests are carried out on materials, components and the final product using two main methods.

- **Non-destructive** testing (or failure testing), where the product is tested until it shows signs of failing, e.g. cracking, to determine how much force is needed to deform it.

Figure 2.48 *Compressive testing of a polymer used in packaging.*

- **Testing to destruction**, where the product is destroyed under controlled conditions and monitored to gather valuable research data, e.g. Euro New Car Assessment Programme (NCAP) for car safety testing.

Quality control in print runs

Paper is not an inert material – it reacts to changes in the environment, which may cause problems during a print run. Paper is affected dramatically by temperature and changes in humidity, which can cause it to curl. The relative humidity of the print room has to be controlled to ensure curl stability, which could affect the colour registration of the printed materials. Usually, when paper stock arrives at the print shop, it is kept in storage for a while in order for it to adapt to the relative humidity of the print room. During the print run itself, regular QC are made to ensure the quality of the printed materials.

Table 2.18 Quality control checks used during a print run.

Problem	Description	Quality control
Set off	The ink from one sheet smudges onto the underside of the following sheet.	Use of sufficient anti-set-off spray, sufficiently quick-drying inks or better quality paper stock.
Colour variation	The printer does not maintain consistent colour throughout the run.	Use of colour bars and regular densitometer readings.
Hickies	Small areas of unwanted solid colour surrounded by an unprinted 'halo' area; caused by specks of dirt, paper debris or ink skin on the printing plate or blanket cylinders.	Regular washing of the blanket cylinder.
Bad register	Colours protrude beyond the edge of the four colour separations, making the image look out of focus.	Regular inspection of registration marks to line up the four colour separations exactly.

Printers' marks

To aid the QC in a print run, printers' marks are used. Colour bars, for example, provide vital information about the performance of both the printing press and the inks being used. They contain a whole range of tests; some are visual checks made by the operator of the printing press, but others can be performed electronically using a densitometer. This instrument monitors the thickness or density of the ink printed on the colour bar to ensure that it is of a consistent quality throughout a print run.

Registration marks help to align the four process colours (CMYK) to form the full colour image. In offset lithography, for example, the full colour image is separated into the four process colours, a printing plate made for each colour and each print unit lays down one of the four process colours at a time. However, the process can be prone to error due to the incorrect registration of one of the printing plates. This results in a low-quality image where one colour is printed slightly 'off centre', causing blurring.

> ### THINK ABOUT THIS!
>
> What would be the consequences of manufacturing a product, such as a magazine, without applying QC procedures?

Figure 2.49 Title page of this book showing the printers' marks.

Total quality management

Total quality management (TQM), often referred to as total quality control (TQC), is the strategic integrated system for achieving customer satisfaction by applying QA procedures at every stage of the production process. TQM is based on all members of an organisation participating in the continual improvement of processes, products, services and the overall culture in which they work. Each department in a company is treated as a client, therefore ensuring high standards of service and attention to detail when dealing between departments. For example, a production team must produce a high-quality component that the assembly team know is quality assured and will therefore fit perfectly.

The British Standards Institute (BSI) operates a quality management system by which any organisation can be accredited to help them produce products of a consistent high quality. Known as the ISO 9000 series of standards, it is the world's most established quality framework, currently being used by over three-quarters of a million organisations in 161 countries. If an organisation is accredited with this standard, the customer is assured of the quality of the product and service.

Figure 2.50 *The BSI Kitemark logo displayed by organisations gaining ISO 9000 accreditation.*

Table 2.19 *Benefits of the BS EN ISO 9000 series of standards.*

Sector	Benefits
Customers and users	Receive products that: • conform to the requirements • are dependable and reliable • are available when needed • are maintainable.
People in the organisation	• Better working conditions. • Increased job satisfaction. • Improved health and safety. • Improved morale.
Owners and investors	• Increased return on investment. • Improved operational results. • Increased market share. • Increased profits.
Society	• Fulfilment of legal and regulatory requirements. • Improved health and safety. • Reduced environmental impact. • Increased security.

2 Quality standards

External formal standards are often used when testing, inspecting and verifying the overall quality of materials, components, products and systems. Formal standards are produced through standards organisations for national, European or international use.

- **National Standards,** e.g. British Standards (prefix BS), are produced by a country's national standards body (NSB). In the UK, British Standards are developed together with the UK government, businesses and society. Some are enforced by regulation, but most standards are voluntary.
- **European Standards (prefix EN)** are produced by the European Committee for Standardisation (CEN), whose members are the NSBs of the European Union countries. In the UK, they are adopted as British Standards (BS EN).
- **International Standards (prefix ISO)** are produced by the International Organization for Standardisation (ISO), whose members are the NSBs of countries all over the world. BSI is a leading member of ISO and represents the UK's interest in the development of international standards. BSI also decides which international Standards to adopt as British Standards (BS ISO).

Kitemark and CE marking

The Kitemark symbol (Figure 2.50) was originally only used in the UK, but it is now recognised throughout the world as a mark of quality. Having a Kitemark associated with a product or service certifies that it complies with a particular standard. The letters 'CE' on a product are the manufacturer's claim that the product meets the requirements of all relevant European Directives. Many products are covered by these directives and for some, such as toys, it is a legal requirement to have a CE mark. This shows that the product achieves a minimum level of quality, and ensures it can be moved freely throughout the European Single Market.

Figure 2.51 *European CE marking for quality. accreditation.*

WEBLINKS:

www.bsieducation.org
www.bsi-global.com

Health and safety

Getting started!

Health and safety is a very important subject. Employers are legally required to minimise the risks to their employees and in turn employees need to take reasonable care when carrying out their jobs. The school D&T department is no different. What would happen if there were no signs on machinery or you were simply allowed to 'do what you want' in a workshop? Acting in an irresponsible way that might cause an injury or illness to yourself or someone else is a criminal offence that might lead to prosecution.

1 Health and Safety at Work Act (1974)

Under this Act of Parliament, employers are legally required to do all that is reasonably practicable to ensure the health, safety and welfare at work of employees, and the health and safety of non-employees such as students and visitors to a school. The following regulations are procedures to safeguard the risk of injury to people.

Personal protective equipment

The *Personal Protective Equipment at Work Regulations, 1992* state that employers have basic duties concerning the provision and use of personal protective equipment (PPE) at work. PPE is defined in the regulations as "all equipment (including clothing providing protection against the weather) which is intended to be worn or held by a person at work and which protects him against one or more risks to his health or safety." These can include safety helmets, gloves, eye protection, high-visibility clothing, safety footwear and face masks or respirators.

The main requirement of the regulations is that PPE is to be supplied and used at work wherever there are risks to health and safety that cannot be adequately controlled in other ways. The Regulations also require that PPE is:
- properly assessed before use to ensure it is suitable
- maintained and stored properly
- provided with instructions on how to use it safely

- used correctly by employees.

Table 2.20 *Hazards and types of personal protective equipment.*

Risk	Hazards	Personal protective equipment (PPE)
Eyes	Chemical or metal splash, dust, projectiles, gas and vapour, radiation.	Safety spectacles, goggles, face shields, visors.
Head	Impact from falling or flying objects, risk of head bumping, hair entanglement.	A range of helmets and bump caps.
Breathing	Dust, vapour, gas, oxygen-deficient atmospheres.	Disposable filtering face-piece or respirator, half- or full-face respirators, air-fed helmets, breathing apparatus.
Protecting the body	Temperature extremes, adverse weather, chemical or metal splash, spray from pressure leaks or spray guns, impact or penetration, contaminated dust, excessive wear or entanglement of own clothing.	Conventional or disposable overalls, boiler suits, specialist protective clothing, e.g. chain-mail aprons, high-visibility clothing.
Hands and arms	Abrasion, temperature extremes, cuts and punctures, impact, chemicals, electric shock, skin infection, disease or contamination.	Gloves, gauntlets, mitts, wrist-cuffs, armlets.
Feet and legs	Wet, electrostatic build-up, slipping, cuts and punctures, falling objects, metal and chemical splash, abrasion.	Safety boots and shoes with protective toe caps and penetration-resistant mid-sole, gaiters, leggings, spats.

Signage

The *Safety Signs (Signs and Signals) Regulations, 1996* require employers to display an appropriate safety sign and instruction wherever a significant risk or harm cannot be avoided or reduced by other means. These Regulations bring into force a European Directive whose purpose is to encourage the standardisation of safety signs throughout Europe so that safety signs, wherever they are seen, have the same meaning.

The Regulations cover various means of communicating health and safety information. These include the use of illuminated signs, acoustic signals such as fire alarms, and traditional signboards such as prohibition, warning and fire safety signs, e.g. signs for fire exits and fire-fighting equipment.

Table 2.21 *Standard health and safety signage.*

Health and safety signage		Example
	Prohibition signs are used to prohibit actions to prevent personal injury and the risk of fire.	Authorised persons only
	Mandatory signs convey action that must be taken e.g. procedures in case of fire.	Eye protection must be worn in this area
	Warning signs are to warn personnel of possible dangers in the work place.	CAUTION Cleaning in progress
	Safe condition signs show directions to areas of safety and medical assistance.	First aid box
	Fire equipment signs show the location of fire equipment and compliance with Fire Precautions.	Fire extinguisher

(Reference: http://www.archersafetysigns.co.uk)

Warning symbols

Warning symbols are placed on products to provide health and safety information for the consumer. For example, British Standard (BS) EN 71 is concerned with the safety of toys, of which Part 6: Graphical Symbol for Age Warning Labelling, covers age warning symbol labelling and specifies the requirements of the symbols used on toys not suitable for children under the age of three.

Many warning symbols appear on the packaging of adhesives and domestic cleaning products along with additional safety instructions that outline any potential risks to users.

Not suitable for children under 3 years (36 months) Irritant to eyes and skin

Figure 2.52 *Warning symbols on packaging.*

WEBLINKS:

www.bsieducation.org – British Standards Institute (BSI)
www.btha.co.uk – British Toy and Hobby Association

THINK ABOUT THIS!

Why is it important for the packaging of toys to carry an age warning symbol? What are the risks to a young child's health and safety by not paying attention to such warnings?

Health and Safety Executive risk assessments

Government guidelines for health and safety issues within the workplace, including schools, are laid out by the Health and Safety Executive (HSE). The HSE states that all places of work must carry out risk assessments of their facilities to identify any potential hazards to employees or students, and to put in place control measures to reduce the risk of injury.

The HSE outlines its 'Five Steps to Risk Assessment':
1. Identify the hazard.
2. Identify the people at risk.
3. Evaluate the risks.
4. Decide upon suitable control measures.
5. Record risk assessment.

WEBLINKS:

www.hse.gov.uk – The Health and Safety Commission is responsible for health and safety regulation in Great Britain.

Using computers

Computers are often used to design graphic products. This will involve large amounts of time sitting at a computer workstation, looking at a monitor or screen, typing at a keyboard and using a mouse – all of which are potential hazards. The HSE, together with the European Union 'VDU Directive', has regulations and guidance on working at a computer workstation.

Repetitive strain injury

Repetitive strain injury (RSI) is a medical condition affecting muscles, tendons and nerves in the arms and upper back. It occurs when muscles in these areas are kept tense for very long periods of time, due to poor posture and/or repetitive motions. There are a number of ergonomic products available to reduce the risks occurring from prolonged

Hazard	Risk	Control measure
Potential (of risk) from a substance, machine or operation.	Reality (of harm from the hazard).	Action taken to minimise the risks to people.

Figure 2.53 *What is the difference between a hazard and a risk?*

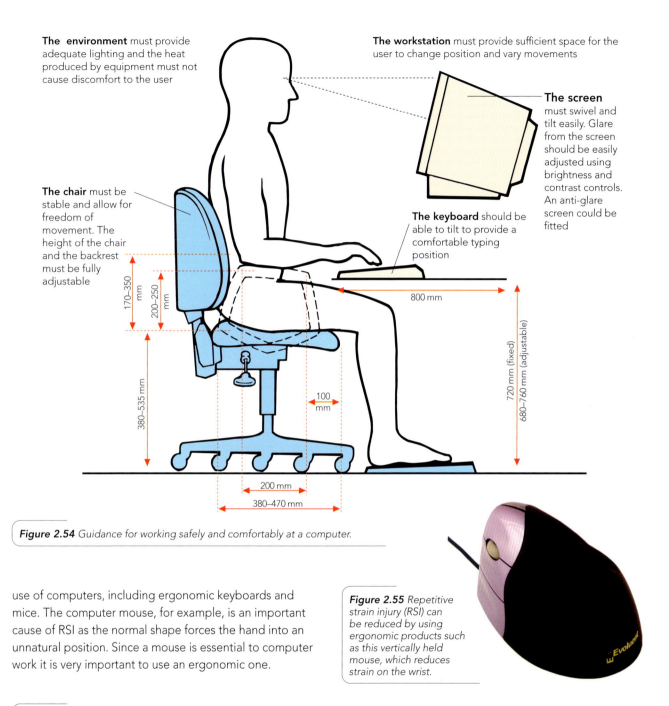

The environment must provide adequate lighting and the heat produced by equipment must not cause discomfort to the user

The workstation must provide sufficient space for the user to change position and vary movements

The screen must swivel and tilt easily. Glare from the screen should be easily adjusted using brightness and contrast controls. An anti-glare screen could be fitted

The chair must be stable and allow for freedom of movement. The height of the chair and the backrest must be fully adjustable

The keyboard should be able to tilt to provide a comfortable typing position

170–350 mm
200–250 mm
380–535 mm
800 mm
100 mm
200 mm
380–470 mm
720 mm (fixed)
680–760 mm (adjustable)

Figure 2.54 *Guidance for working safely and comfortably at a computer.*

use of computers, including ergonomic keyboards and mice. The computer mouse, for example, is an important cause of RSI as the normal shape forces the hand into an unnatural position. Since a mouse is essential to computer work it is very important to use an ergonomic one.

Figure 2.55 *Repetitive strain injury (RSI) can be reduced by using ergonomic products such as this vertically held mouse, which reduces strain on the wrist.*

Table 2.22 *Part of a risk assessment for using a computer.*

Hazard	Risk	People at risk	Control measure
Using a computer	Repetitive strain injury (RSI)	User	• Keyboard should tilt to provide a comfortable typing position. • Use an ergonomic keyboard with wrist support. • Use an ergonomic mouse. • Take regular breaks to rest hands.
	Eye strain	User	• Adjust glare from monitor using brightness and contrast controls. • Use of an anti-glare screen fitted to monitor to reduce screen flicker. • Tilt or swivel monitor to reduce reflections. • Take regular breaks to rest eyes.

Table 2.23 Part of a risk assessment for using a pillar drill.

Hazard	Risk	People at risk	Control measure
Using a pillar drill	Damage to eyes from flying debris	User/people in immediate area	• Use appropriate PPE i.e. safety specs or goggles. • User fully briefed on use of machine (general machine safety), i.e. guards in position. • Appropriate supervision by teacher or technician. • Other students wait behind marked yellow lines or barriers when not using machine.
	Cuts from metal shavings	User	• Work clamped securely in vice (never held in hand) to prevent work from catching and spinning. • Use of appropriate PPE, i.e. gloves. • Use a small 'stick' to remove large spiral shavings. • Place shavings into appropriate disposal container.

Workshop practices

When making models and prototypes in Graphic Products it is essential that a fully equipped workshop is used. Obviously, there are a wide range of potential hazards when using machinery, power tools and equipment. The school or college will have carried out a detailed risk assessment for each piece of equipment, which should be clearly displayed for your information.

THINK ABOUT THIS!

In small groups, carry out a series of risk assessments for different workshop machinery, equipment and processes. Collate your work to build up a comprehensive set of risk assessments that you can use in your coursework projects.

Control of Substances Hazardous to Health (COSHH) regulations

These regulations place a duty on employers to make an assessment of risks for work involving exposure to substances hazardous to health. Steps must be taken to prevent or control adequately the exposure of employees and others to these substances. Hazardous substances include:

- substances used directly in work activities such as adhesives, paints and cleaning agents

- substances generated during work activities such as fumes from soldering and welding

- naturally occurring substances such as dust

- biological agents such as bacteria and other micro-organisms.

Table 2.24 Health and Safety Executive (HSE) guidance on COSHH regulations.

Step		Action
1	Assess the risks	Assess the risks to health from hazardous substances used in or created by workplace activities.
2	Decide what precautions are needed	Do not carry out work that could expose employees to hazardous substances without first considering the risks and the necessary precautions, and what else is needed to comply with COSHH.
3	Prevent or adequately control exposure	Prevent employees being exposed to hazardous substances. Where preventing exposure is not reasonably practicable, then it must be adequately controlled.
4	Ensure that control measures are used and maintained	Ensure that control measures are used and maintained properly and that safety procedures are followed.
5	Monitor exposure	Monitor the exposure of employees to hazardous substances, if necessary.

6	Carry out appropriate health surveillance	Carry out appropriate health surveillance where assessment has shown it necessary or where COSHH sets specific requirements.
7	Prepare plans and procedures to deal with accidents, incidents and emergencies	Prepare plans and procedures to deal with accidents, incidents and emergencies involving hazardous substances, where necessary.
8	Ensure employees are properly informed, trained and supervised	Provide employees with suitable and sufficient information, instruction and training.

Many adhesives are solvent-based, containing volatile organic compounds (VOCs) that give off vapours that can cause dizziness and nausea. Because of this, these substances are extremely hazardous to use within confined indoor areas such as workshops or classrooms. It is important that thorough risk assessments are carried out and the appropriate action taken to minimise the risks.

WEBLINKS:

www.coshh-essentials.org.uk – COSHH Regulations

Table 2.25 *Part of a risk assessment for storage and use of solvent-based adhesives.*

Hazard	Risk	People at risk	Control measure
Use of solvent-based adhesives	Burns from corrosive adhesives	User	• Use appropriate PPE including gloves and eye protection. • Users fully briefed on safe use of adhesives. • Appropriate supervision by teacher or technician. • Wash area immediately with warm soapy water – seek medical attention. • Eyes – seek medical attention immediately; use an eye bath to wash eyes.
	Inhalation of VOC vapours	User/people in immediate area	• Use only in well-ventilated areas, i.e. use extraction or open external windows/doors. • Appropriate supervision by teacher or technician. • Use of face-mask or respirator. • If dizziness and nausea occur – vacate area immediately and seek medical attention.
	Storage	Technician and teaching staff	• Store in a secure metal cupboard. • Cupboard easily identifiable (yellow) with appropriate safety signage clearly displayed. • Staff to be fully briefed as to safe storage of adhesives. • Checks by technician on a regular basis.

ExamCafé

Relax and prepare

SURPRISE

Revision Summary

It is important to note that the questions asked by the examiner in this exam paper will cover aspects from all four sections of this unit. No paper will ever focus upon one section entirely. Therefore, it is vital that you have a secure knowledge and understanding across all four sections.

You should give yourself plenty of opportunities to answer examination-style questions throughout the course so you are prepared for the final examination. Use the sample assessment materials (SAMs) and past exam papers provided by Edexcel (www. edexcel.org.uk) and the questions in this textbook.

Don't forget – if in doubt, ASK! Your teacher is there to help you understand the theory in this unit. Have a good, long think about appropriate questions to ask your teacher – it might be a good idea to discuss it with your peers first to see if they can explain it more clearly.

Finally, keep a set of well-ordered and legible revision notes, which will help you to learn key topics and provide you with something to refer back to when in doubt.

Refresh your memory

Revision checklist

▷ Make sure that you have answered all the questions at the end of this section.

▷ Make sure that your revision notes are well ordered, clear and up-to-date.

▷ Use the web links to read around each key topic so that you are well informed.

▷ Use sample assessment materials and past papers to practice your exam technique.

▷ Discuss any problems with your peers or teacher – don't keep them to yourself!

Get the result !

Tips for answering questions

Always read each question carefully before you respond. It might be a good idea to use a piece of scrap paper to outline your response if you think you have enough time.

Always look at the amount of marks awarded for each question in brackets. This will give you a good indication of how many points need to be raised in your response. As a general rule of thumb, look at the following command words and what you have to do in order to gain the marks:

Give, State, Name	(1 mark)	These types of questions will not feature heavily at A2 level but may appear at the beginning of the paper or question part. They are designed to ease you into the question with a simple statement or short phrase.
Describe, Outline	(2+ marks)	These types of questions ask you to simply describe something in detail. Some questions may also ask you to use notes and sketches; you can gain marks with the use of a clearly labelled sketch.
Explain, Justify	(2+ marks)	These types of questions will be commonplace in this exam. They are asking you to respond in detail to the question – no short phrases will be acceptable here. Instead, you will have to make a valid point and justify it.
Evaluate	(4+ marks)	These types of questions will appear towards the end of the paper or question part and are designed to stretch and challenge the more able student. They require you to make a well-balanced argument, usually involving both advantages and disadvantages.

Model answers

The following four questions should demonstrate the style of questions using the different types of command words. The places where marks have been awarded are indicated in brackets. These are referred to as 'trigger points' and are parts of the examiner's mark scheme where it is expected marks will be awarded.

Exam question 1

Give **two** reasons why a thin layer of tin is added to a steel drinks can. **(2)**

This question is a straightforward 'give' question, so short statements are acceptable and they do not have to be justified.

Danny

1. *Stops can from rusting* **(1)**
2.

   (1 mark)

Here, the student has given one relevant reason for using tin. However, they have not even attempted a second response. It is really important that you attempt all questions – even if you have to make an educated guess!

Emma

1. *Tin prevents corrosion of the steel can* **(1)**
   ....
2. *Protective layer extends the shelf-life of the product* **(1)**

   ....
   (2 marks)

Two well constructed sentences that give two appropriate reasons for using a layer of tin on a steel drinks can – full marks.

Exam question 2

Explain **two** reasons why polystyrene (PS) is used for the casing of many electrical products, such as mobile phones. **(4)**

This question asks you to apply your knowledge and understanding of polymers (polystyrene, to be specific) to a familiar product. In this question you are asked for two explanations worth two marks each. In other words, two fully justified points.

Louise

1. *You can get polystyrene in a wide range of colours* **(1)** *so your mobile phone can be different colours.*
2. *Polystyrene is lightweight.* **(1)**
   (2 marks)

Here, the student has given two acceptable properties of polystyrene that make it suitable for use in a mobile phone. However, the first answer is not fully justified and the second offers no justification at all.

Sally

1. *Polystyrene is a thermoplastic,* **(1)** *which means that it can be easily formed* **(1)** *into a casing using injection moulding.*
2. *Polystyrene is lightweight,* **(1)** *so it is ideal for portable devices* **(1)** *such as mobile phones.*
   (4 marks)

This student achieves full marks for this question as both properties of polystyrene are fully justified when applied to its use in mobile phones.

Exam question 3

Justify the use of a forehead thermometer with thermochromic liquid crystals rather than the traditional glass and mercury thermometer when taking the temperature of a young child. **(4)**

This question relates to smart materials and in particular thermochromic liquid crystals. You must apply your knowledge and understanding of thermochromic liquid crystals to their application as a forehead thermometer and then compare it with another similar product. Again you are asked for two fully justified points, although this time the question is not as structured as 2 x 2 marks. Instead, the mark in brackets indicates that there should be four marking points.

Paul

They are safer **(1)** *and they are cheaper than traditional thermometers.* **(1 mark)**

Here, the student has given one relevant point but has offered no justification as to why thermochromic liquid crystal thermometers are safer. The second point is not relevant to this question. They may actually be 'cheaper', but what has this fact got to do with taking the temperature of a young child?

Adam

They are safer **(1)** *because they will not crack like glass or cause mercury poisoning.* **(1)** *You can also take the temperature on the child's forehead rather than in the mouth,* **(1)** *which is much less stressful for the child.* **(1)**
(4 marks)

This response gets full marks because each point is relevant and fully justified.

Exam question 4

Evaluate the use of mass production to manufacture many consumer goods. **(6)**

This question asks for an 'evaluation' of a topic so the response must look at both sides of the argument – using both advantages and disadvantages. It is a type of question that is supposed to stretch students because it is more open ended.

Lynsey
Mass production is a good way of producing thousands of products because people want to buy them so they have to cater for the demand **(1).** *When thousands are produced the cost of manufacture reduces so products are cheaper* **(1).** *Products aren't as dear so more people will buy them and that's why mass production is good.*
(2 marks)

Here, the student has started well with two relevant points (although not really justified enough to gain two marks each) but soon runs out of steam. The

last sentence repeats an earlier point – 'products are cheaper' is the same as 'products aren't as dear'.

Barbara
Mass production is a suitable scale of production for many consumer goods because it takes advantage of economies of scale **(1)** *and buying materials and components in bulk making them cheaper.* **(1)** *Mass production uses automated machinery on production lines,* **(1)** *which is a highly efficient* **(1)** *way of manufacturing on a large scale.*

However, mass production can have many negative effects upon the workforce. Many workers are low skilled **(1)** *and simply mind machines. This can lead to poor job satisfaction* **(1)** *due to the mundane and repetitive nature of the job.*
(6 marks)

This response gains full marks as it addresses both the advantages and disadvantages of mass production. The response is succinct and well structured, containing six trigger points from the examiner's mark scheme.

Practice questions

1. Packaging materials are used to protect televisions while in transit.

   (i) Name the most suitable type of carton board for the outer packaging. **(1)**

   (ii) Name a suitable polymer for use in the internal packaging. **(1)**

2. Give **one** advantage and **one** disadvantage of using mechanical wood pulp in the production of paper and board. **(2)**

3. Medium-density fibreboard (MDF) and Styrofoam™ are often used to produce prototype models. Explain **two** advantages of using Styrofoam to model a product at the development stage. **(4)**

4. Explain **two** reasons why PET is used in the manufacture of a fizzy drinks bottle. **(4)**

5. Outline the following stages in the commercial production of a carton board package.

   (i) Preparation **(2)**

   (ii) Processing/production **(4)**

   (iii) Assembly **(2)**

   (iv) Finishing **(2)**

6. Sketch a 3rd angle orthographic drawing of the metal component below. The front and side elevations and the plan are labelled for you.

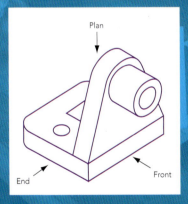

Plan
End
Front
(6)

7. Justify the use of quality control (QC) in the manufacturing process. **(4)**

8. Explain how a designer could minimise the risk of repetitive strain injury (RSI) when using a computer for prolonged periods of time. **(4)**

9. Evaluate the use of one-off production to produce products. **(6)**

10. Evaluate the widespread use of polymers to produce commercial packaging. **(6)**

Design for the Future

Summary of expectations

1 What to expect

In this unit, you will develop your knowledge and understanding of a range of modern design and manufacturing practices and contemporary design issues. The modern designer must have a good working knowledge of the use of information and communication technology (ICT) and systems and control technology in the design and manufacture of products. They must also be aware of the important contributions of designers from the past, which may provide inspiration for future design.

It is increasingly important that you develop an awareness of the impact of design and technological activities upon the environment. Sustainable product design is a key feature of modern design practices.

2 How will it be assessed?

Your knowledge and understanding of topics in this unit will be externally assessed through a 2-hour examination paper set and marked by Edexcel. The exam paper will be in the form of a question and answer booklet consisting of short-answer and extended-writing type questions.

The total number of marks for the paper is 70.

3 What will be assessed?

This unit is divided into four main sections, with each section outlining the specific knowledge and understanding that you need to learn:

3.1 Industrial and commercial practice
- Information and communication technology (ICT).
- Digital special effects.
- Biotechnology.

3.2 Systems and control
- Manufacturing systems.
- Computer-integrated manufacture (CIM).
- Robotics and Artificial Intelligence (AI).
- Flow charts.

3.3 Design in context
- The effects of technological changes on society.
- Influences of design history on the development of products.
- Form and function.
- Anthropometrics and ergonomics.

3.4 Sustainability
- Life cycle assessment (LCA).
- Cleaner design and technology.
- Minimising waste production.
- Renewable and non-renewable sources of energy.
- Responsibilities of developed countries.

4 How to be successful in this unit

To be successful in this unit you will need to:
- have a clear understanding of the topics covered in this unit
- apply your knowledge and understanding to a given situation or context
- use specialist technical terminology where appropriate
- write clear and well-structured answers to the exam questions that target the amount of marks available.

5 How much is it worth?

This unit is worth 40 per cent of the A2 course and 20 per cent of the overall full Advanced GCE.

Unit 3	Weighting
A2 level	40%
Full GCE	20%

Industrial and commercial practice

Getting started!

This section follows on from **Unit 2: Industrial and commercial practice** and develops your knowledge and understanding of modern design practice. We all know that the modern world relies upon computers and that the modern workforce must be computer literate – but what are the advantages and disadvantages of using these systems? Biotechnology is a controversial new technology where living organisms are genetically modified to produce products for a specific use – is this right? What impact do these modern technologies have upon our lives?

1 Information and communication technology

Electronic communications

E-mail

Information and communication technology (ICT) has improved the reach (level of communication across a network) and range (types of data transfer available) of electronic communications. E-mail is the simplest form of electronic communication and has a comparatively low level of reach and range when it is used for messaging or transferring documents. However, it has proved invaluable in rapid communications between designers, manufacturers, retailers and consumers due to its ease of use and widespread access through Internet connections. For these reasons it has all but replaced the postal system. There are issues of security and privacy when using e-mail

and limitations on the size of attachments, which often restrict its use, but to the majority of people it is now their preferred way of communicating.

Electronic data interchange

Electronic data interchange (EDI) is a new way for companies to do 'paperless' business using a process that transfers business documents through a computer network, rather than via the postal system. Many modern companies use EDI as a fast, inexpensive and secure system of sending purchase orders, invoices, design and manufacturing data, etc. For example, some manufacturers use EDI to transmit large, complex CAD drawings and multinational companies use EDI to communicate between locations world-wide. EDI can also be used to transmit financial information and payment in electronic form. However, the transfer of files requires that the sender and receiver agree upon a standard document format for the document that is to be transmitted.

The EDI process starts with a trading agreement between a company and their trading partner. Joint decisions have to be made regarding the standard to be used, the information to be exchanged, how the information is to be sent, and when information will be sent. To send a document, EDI translation software is used to convert the document format into the agreed standard. The translator creates and wraps the document in an electronic envelope with an identification code and sends it to the partner's mailbox. The document is retrieved from their mailbox and an EDI translator opens the envelope and translates the data from the standard form to their application's format. The translator ensures that the data sent by one company is converted into a format that another can use.

Table 3.1 Advantages and disadvantages of e-mail.

Advantages	Disadvantages
• Quick, easy and convenient means of communicating around the world. • Widespread usage (anyone with a computer connected to the Internet). • E-mail exchanges can be saved as a dated record of correspondence. • Documents can be attached electronically and can be saved and edited easily.	• Impersonal and some messages can be misinterpreted. • Influx of messages to inbox increases time to read and respond. • 'Spamming' of unsolicited commercial e-mails, often with inappropriate content. • Privacy and security issues as messages can be intercepted and read. • Limitations on size of attachments.

Table 3.2 *Advantages and disadvantages of EDI for business.*

Advantages	Disadvantages
• Saves money by eliminating the need for processing paper documents. • Saves time as information is transferred digitally. • Improves customer service as business documents are transferred quickly with fewer errors. • Expands customer base due to improved customer service through efficient EDI processes.	• Incompatibility with some companies due to range of standard document formats. • Standards updated regularly, which causes problems with different versions in use. • Expensive to initially set up. • Limits trading to only companies with EDI.

Integrated Services Digital Networks and broadband

The development of Integrated Services Digital Networks (ISDN) and, more recently, broadband technology, means that huge amounts of information can be transferred across computer networks at far greater speeds than ever before. The purpose of ISDN is to provide fully integrated digital services to users comprising digital telephony and data-transport services through existing telephone networks. ISDN involves the digitisation of the telephone network, which enables voice, data, text, graphics, music and video to be transmitted at high speeds over existing telephone lines. The emergence of ISDN represented an effort to combine subscribed services such as Internet service provision and telecommunications into one package. It enabled users to have additional phone lines installed so that the Internet could be used at the same time as the telephone without callers receiving a 'busy' signal.

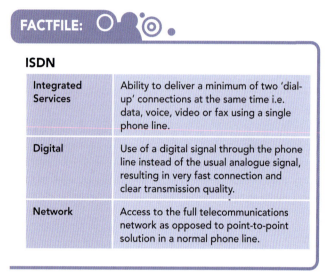

FACTFILE:

ISDN

Integrated Services	Ability to deliver a minimum of two 'dial-up' connections at the same time i.e. data, voice, video or fax using a single phone line.
Digital	Use of a digital signal through the phone line instead of the usual analogue signal, resulting in very fast connection and clear transmission quality.
Network	Access to the full telecommunications network as opposed to point-to-point solution in a normal phone line.

Recent developments in digital technology have seen broadband becoming one of the fastest-growing new consumer technologies. There are many broadband service providers currently competing for this expanding market in the UK alone. Between them they offer five types of broadband access.

Table 3.3 *Comparison of different types of broadband access.*

Type of broadband	Description	Advantages	Disadvantages
Asymmetric digital subscriber line (ADSL)	The most popular type of broadband in the UK. It is a way of sending data along an existing telephone line using technology that packs more information into the signal.	• Easy to install.	• Downloading information is much faster than sending.
Cable	The cable itself is made up of a bundle of optical fibres that can carry many times more information than a telephone cable and over longer distances without the signal quality decreasing.	• Easy to install. • Potentially the fastest connection.	• Only available in cities and large towns.
Wireless	Wireless, or 'Wi-Fi®', connections are made by transmitting the Internet to a user's computer by means of radio signals. Wireless local area networks (LANs) can be built by putting aerials on the outside of a number of houses which then allow anyone within a certain area to receive broadband Internet access.	• Not restricted by the local telephone exchange. • Fast connection for downloading and sending information.	• Requires an experienced technology provider to set up and maintain the network. • Can suffer from interference in built-up areas if there is another network nearby used for something other than Internet access.

3G technology	Offered by mobile phone networks allowing mobile phones, 'BlackBerrys®' and laptops to connect to the Internet wherever they are, as long as there is sufficient reception.	• Internet connectivity on the move.	• Restrictions to the attachments that can be opened due to software limitations.
Satellite	There are two types of broadband available by satellite: one-way and two-way. One-way provides a very fast connection but in one direction – download only. This means to send information back a dial-up connection is still needed. Two-way satellite allows the ability to both download data and send it back to the satellite – though at a much slower rate.	• Available anywhere.	• Reasonably expensive. • Bad weather can cause connection problems.

Videoconferencing

ISDN and broadband have enabled high-speed data and communications transfer, which can be used in a videoconferencing system. A videoconference allows two or more locations to interact using two-way video and audio transmissions simultaneously, enhancing communications and speeding up the decision-making process by eliminating the need for time-consuming travel to meetings, which might be across the other side of the world.

A videoconferencing system includes a video camera to capture images, a screen to view images, microphones to pick up sound and loudspeakers to play sounds. Data is transferred via the Internet using ISDN or broadband. There are two types of videoconferencing systems.

- **Dedicated systems** have all the required components packaged into a single console, including a high-quality remote-controlled video camera.

- **Desktop systems** are add-ons to normal personal computers such as webcams and microphones, transforming them into videoconferencing devices.

Multipoint videoconferencing allows for simultaneous videoconferencing among three or more remote points using a multipoint control unit (MCU) as a bridge that interconnects calls from several sources. This enables three or more people to sit in a 'virtual' conference room and communicate as if they were sitting next to each other.

Table 3.4 Advantages and disadvantages of videoconferencing.

Advantages	Disadvantages
• Eliminates the need for travel to other countries, saving both time and money. • Visual information can be communicated as part of the conversation. • Accelerates the decision-making process as presentations can be made to several people simultaneously. • Remote diagnostics available with technicians in other countries able to solve problems. • Corporate training of many staff at the same time.	• Synchronisation of time of meeting in different times zones across the world. • Connection can often fail. • Lack of eye contact with others in meeting can hinder conversation or intent. • 'Camera shyness' can hinder presentations due to pressures of being filmed and often recorded.

Figure 3.1 Multipoint videoconferencing allows people in three or more countries to communicate with each other simultaneously.

THINK ABOUT THIS!

Is ICT bringing us closer together? Granted, we can communicate quickly over vast distances and whilst on the move, but have we lost the personal touch? Is is too convenient to e-mail someone, which is quite anonymous, instead of speaking to them face-to-face?

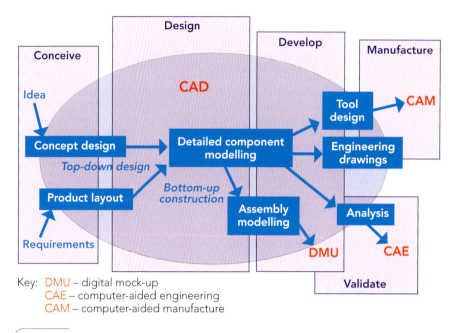

Key: DMU – digital mock-up
CAE – computer-aided engineering
CAM – computer-aided manufacture

Figure 3.2 *CAD in the design process.*

Computer-aided design (CAD)

In fast-moving sectors like graphic and product design, the deadlines set by clients are often incredibly tight. This creates considerable pressure on the design process to get a product to market on time and on budget. Compared with traditional drafting techniques, manufacturers can operate more efficiently using a computer-aided design (CAD) system as they can create better designs faster and at a lower cost. The introduction of CAD/computer-aided manufacturing (CAM) techniques means CAD data is now employed to automatically generate tool paths for automated machine tools. The application of CAD systems has revolutionised design practice.

Creative and technical design

A CAD system incorporates hardware (computer) and software (CAD program), enabling designers to work individually or in teams to design products. CAD provides increased flexibility for designers and allows them greater control over the quality of the finished product. CAD is used throughout the design process, starting with conceptual designs that can be quickly edited and modified with client feedback. Detailed engineering drawings can be generated from 3D models that provide vital manufacturing information. Computer modelling can test components on-screen and the creation of photorealistic images is ideal for marketing use. It is commercially beneficial to associate CAD with CAM where

possible. CAM software uses the geometric data from the CAD program to generate instructions to drive automated machine tools, such as computer numerically controlled (CNC) lathes. As a final check, prior to manufacture, it is possible to upload the program back into the CAD system for a path-trace of the cutting tool.

CAD systems reduce the need for large and labour-intensive drawing offices, which was the traditional means of producing technical drawings. Designers using CAD are multi-skilled, with high levels of computer and visual literacy alongside creativity and problem-solving skills. A designer does not even have to work in the design office as all CAD data can be transferred electronically. Increasingly, CAD files are sent across the world to regions where manufacturing is inexpensive.

The ease with which a designer can use a CAD system to generate complex engineering drawings displays its flexibility as a drafting tool. A CAD system can store individual drawings in libraries allowing the designer to select a wide variety of commonly used components. This improves both the quality and consistency of the drawing and the speed at which it is completed. The many advantages of a CAD system all serve to increase the quality of and reduce the time taken to generate an engineering drawing, incorporating:

- the ability to 'zoom in' and 'zoom out' when drawing to scale

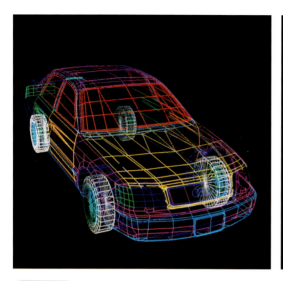

Figure 3.3 *CAD combines creative and technical design features.*

- the use of automatic chamfers, fillet radii and dimensions
- the ability to cut, copy, paste, revolve, rotate and mirror objects
- a wide variety of hatching styles and fonts available
- the ability to easily modify an existing drawing.

Computer-aided engineering (CAE) utilises computer simulation to analyse designs. It can determine, for example, whether assemblies fit together to the required tolerance and whether there is sufficient clearance between moving components. When the product is finalised, test programs can be run to determine cutting tool paths to ensure efficient production with minimum waste. The

Figure 3.4 *Using CAD to test designs ensures a high-quality product.*

THINK ABOUT THIS!

The use of computers is a valuable communication tool for your coursework project. Don't be afraid of using CAD throughout the design process or simply using ICT to enhance the presentation of individual pages. Your entire coursework project can be submitted as an e-portfolio if you are really confident in the use of computers.

Virtual modelling and testing

The ability to represent an accurate 3D model of a product or component on a computer in a 'virtual world' without actually having to build a model is a great advantage to the designer and significantly reduces development times and costs. 3D computer models are used along with prototype models to aid visualisation but have the advantage of being easily changed. A function called 'bi-directional parametric association' aids modifications by causing the slightest change to a design feature to automatically change any other design feature linked with it.

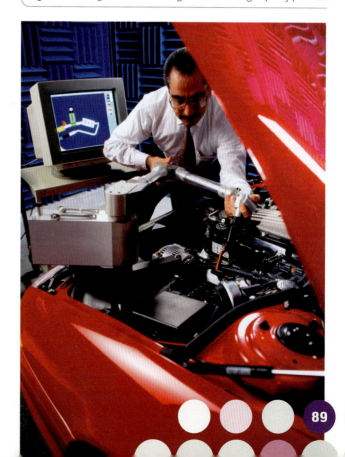

design data collected from these tests can then be directly output to manufacturing facilities. Computer software models are excellent at examining manufacturing systems that are often too complex for humans to understand unless real-life experimentation is used. Examining the manufacturing system in virtual reality has many benefits compared with real-life experimentation: simulations are relatively inexpensive, they can be performed in a relatively short time and they are safe and repeatable.

Rapid prototyping

Models that have historically taken days or weeks to construct through traditional modelling methods can be made in just a few hours using a rapid prototyping (RPT) system such as stereolithography. This allows designers, production directors and marketing personnel the ability to review the function, ease of manufacture and marketability of a product within days of the initial design, shortening design and development cycles, enhancing design quality and accuracy and reducing development costs.

RPT machines have the capability to produce solid models from a variety of materials including numerous plastics, ceramics, wood and metals by taking thin horizontal cross sections from a 3D computer model to construct the physical model layer by layer. Compared with traditional machining methods, which form by-wastage, rapid prototyping offers advantages as complex and intricate models can be produced without the need for complicated machine-tool set-up.

Table 3.5 *Applications and advantages of rapid prototyping.*

Application	Advantages
Prototyping	The primary use is to quickly create prototypes for communication and testing purposes. Prototypes dramatically improve communication as most people find 3D objects easier to understand than CAD drawings. Prototypes are also useful for testing a design to determine whether it performs a certain function.
Rapid tooling	The automatic fabrication of production-quality machine tools using RPT techniques. Tooling is one of the most expensive steps in the manufacturing process because of the extremely high quality required. CNC and manual machining have traditionally been used to make machine tools but these are both expensive and time consuming. Rapid tooling enables the manufacture of standard machine tools in prototype lead times.
Rapid manufacturing	The automated manufacture of saleable products directly from 3D CAD data. Cost effective for low volume and small components where the unit cost is high as it eliminates the need for expensive tooling. It can also provide custom-made components to the customer's exact specifications.

LINKS TO:

Unit 2: Industrial and commercial practice: modelling and prototyping.

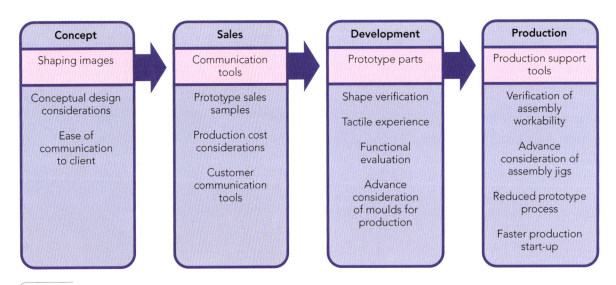

Concept	Sales	Development	Production
Shaping images	Communication tools	Prototype parts	Production support tools
Conceptual design considerations	Prototype sales samples	Shape verification	Verification of assembly workability
Ease of communication to client	Production cost considerations	Tactile experience	Advance consideration of assembly jigs
	Customer communication tools	Functional evaluation	Reduced prototype process
		Advance consideration of moulds for production	Faster production start-up

Figure 3.5 *Rapid prototyping in the product design cycle.*

Marketing, distribution and retail

Electronic point of sale

Information is at the centre of any business and, if used properly, it ensures the business stays one step ahead of its competitors. By using electronic point of sale (EPOS) systems, a business is able to supply and deliver its products and services faster by reducing the time between the placing of an order and the delivery of a product.

Each product can be electronically identified using its unique barcode. When passed over a barcode reader or scanner, the barcode is read by a laser beam. The laser scans the bar code and reflects back on to a photoelectric cell. The bars are detected because they reflect less light than the background on which they are printed. Each product has its own unique 13-digit number. The first two numbers indicate where the product was made, the next five are the brand owner's number, the next five are given by the manufacturer to identify the type of product and the final digit is the check digit, which confirms that the whole number has been scanned correctly.

| 4 0 | 0 2 5 8 9 | 0 0 1 1 8 | 7 |
| Origin of product | Brand owner's number | Item number | Check digit |

Figure 3.6 A standard 13-digit barcode.

It is important to note that information regarding the price of the product is not contained on the barcode. Instead, the scanner, at a supermarket checkout for example, transmits the product code number to an in-store computer that relays the product's description and price back to the checkout, where it is displayed electronically and printed on the till receipt. The in-store computer then deducts the item purchased from the stock list so that it can be re-ordered when stock is low.

Data matrices, also known as 2D barcodes, are visual codes that can be read and decoded by machine vision systems. The increasing use of data matrix codes arises from the

Figure 3.7 Data matrices (2D barcodes) are increasingly used as they can contain more information because they use not only the width of the lines but also the height of them.

manufacturer's requirements for tracking their products or components. The intention is that batch or serial numbers can be permanently marked onto components, which is useful for tracking defective batches and identifying counterfeit parts.

EPOS and the associated management software provides manufacturers with:

- a full and immediate account of the financial transactions involving the company's products
- data that can be input into spreadsheets for sales/profit margin analysis
- the means to monitor the performance of all product lines, which is particularly important in mass production as it allows the company to react quickly to demand
- accurate information for identifying consumer buying trends when making marketing decisions
- a full and responsive stock control system by providing real-time stock updates
- a system that ensures sufficient stock is available to meet customer needs without over-stocking, which ties up capital.

Internet marketing and sales

The development of the Internet as a means of competing in a global marketplace has revolutionised the marketing and sales of products and services. Through the global networking of computers, the Internet provides an effective means of accessing a wealth of information and

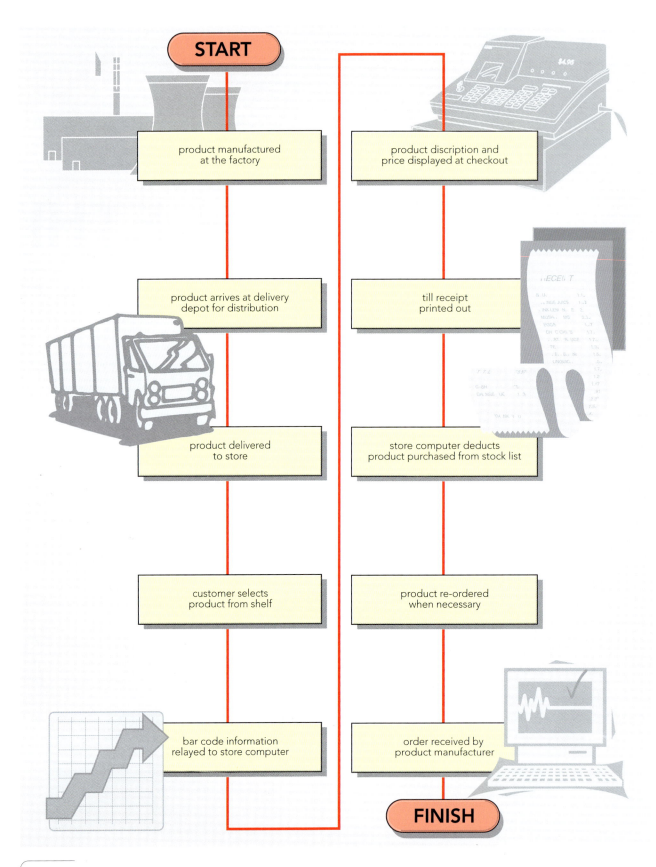

START

product manufactured
at the factory

product discription and
price displayed at checkout

product arrives at delivery
depot for distribution

till receipt
printed out

product delivered
to store

store computer deducts
product purchased from stock list

customer selects
product from shelf

product re-ordered
when necessary

bar code information
relayed to store computer

order received by
product manufacturer

FINISH

Figure 3.8 *The electronic point of sale (EPOS) system.*

entertainment for anyone with a computer connected to the Internet. The dramatic rise in e-commerce has led to virtual communities being formed, which businesses are eager to explore. The possibilities for innovative marketing techniques are endless due to the simple identification of target market groups by user preferences. Marketing can be 'tailor made' to suit these markets, so a marketing message can be sent directly to potential customers as opposed to 'blanket' advertising in traditional media.

Table 3.6 Advantages and disadvantages of Internet sales and marketing.

Advantages	Disadvantages
To manufacturers and retailers: • world-wide reach and access to new markets and increased customer base • increased company profile on a world-wide basis • faster processing of orders and transactions, resulting in efficiency savings and reduced overheads • detailed knowledge of user preferences and market trends by tracking sales • cost savings due to reduced sales force and need for retail outlets • less expensive than traditional advertising media such as TV or magazines • innovative marketing tactics can be employed that target specific groups. To the consumer: • access to a wide range of products and services • availability of product information to inform purchasing decisions • online discounts and savings through price comparison websites • convenience of shopping at home.	• Security concerns regarding input of personal bank details when purchasing goods. • Personal information can be shared with other companies without customer consent. • Difficult to find websites without exact details, resulting in a need for other expensive marketing methods, e.g. magazine adverts, Internet service provider sponsered searches, etc. • Slow Internet connection can cause difficulties in accessing information. • Difficullty in navigating complicated Web pages. • Does not allow 'hands on' experience of product, i.e. touch, taste and fit. • Access to inappropriate material. • Spread of 'junk' mail and threat of computer viruses.

THINK ABOUT THIS!

As a consumer, what are your main reasons for buying products over the Internet? Is it convenience, price, or that you can buy products from other parts of the country or even the world?

2 Digital special effects

Compositing using blue screen and green screen

This is a popular and widely used effect as it enables actors or scale models to be imposed on a separate filmed or computer-generated background. This technique can be used to create scenes that would be impossible or too dangerous to film or to impose actors in a futuristic computer-generated scene. Many 'virtual' news studios are now set up in this manner, with the newsreaders sitting in front of blue screens with headlines and graphics imposed behind them. Weather reporters often use a monitor to the side of the screen to see where they are putting their hands on the imposed weather map.

The background and foreground of a scene are shot separately as two pieces of film and then combined in a process known as compositing. This is achieved using digital technology where the two films are digitised and the composite image made on a computer before being processed as film. First, the background is filmed or created using computer-generated images (CGI) – this is called a background plate. Then the actor is filmed against a blue background or screen, which can be passed through a red filter to make it appear black. Silhouettes or mattes are created of the actor from the blue screen footage; one is black on a white background and the other is white on a black background. There are now four layers of film:

- computer-generated background image – background plate
- film footage of the actor in the foreground
- matte – black silhouette on white background
- matte – white silhouette on black background.

These four layers are combined to make a composite image. The black silhouette is placed on the background plate, creating a 'hole' into which the footage of the actor can be accurately placed. Developments in digital technology have also made it easier to incorporate motion into composited shots, where reference points are placed onto the blue background (usually as a painted grid, 'X's marked with tape, or equally spaced tennis balls attached to the wall). In post-production, a computer can use the references to adjust the position of the background, making it match the movement of the foreground perfectly. Unwanted elements such as safety wires are removed in post-production by 2D painting.

Figure 3.9 *The use of green screen to produce composite images.*

Over the last decade, the use of green screen has become dominant in film special effects. The main reason for this is that green emits more light than blue, making it easier to work with. The choice of colour is up to the effects artist and the needs of the specific shot. Red is usually avoided because it features in human skin tones, but can often be used for objects and scenes that do not involve people.

Computer-generated images

Computer-generated images (CGI) are now commonplace in films, where they are used to create visual effects that would not be possible using traditional means. CGI mixed with live action gives rise to some extremely dramatic effects but many films are now entirely computer generated. The techniques used to manipulate existing images are part of a stage called post-production.

3D modelling is the process of developing a wireframe representation of any 3D object using specialised software. This wireframe is then rendered with both colour and textures and lighting effects added. Textures allow 3D models to look significantly more detailed and realistic than

a smooth plastic look. 3D models can be animated using a technique called 'keyframing', which involves creating key moments of change, known as keyframes. The movement in between these keyframes is calculated by the software to ensure a smooth transition. Often a skeleton is added to the 3D model to make it easier to manipulate movement, and therefore animate. For even greater realism, however, motion capture is used.

Motion capture involves a performer wearing markers near each joint to identify the motion by the positions or angles between the markers. The motion capture computer software tracks the movement of the markers and records the positions, angles and speed to provide an accurate digital representation of the motion. For example, the CGI character Gollum from *The Lord of the Rings* trilogy was created using motion capture from an actor resulting in extremely realistic and natural movements. This technology has also proved useful in sports science for improving the performance of sportsmen and women, for example by tracking the movement of a golfer's swing to diagnose problems with technique.

Figure 3.10 *Creating CGI effects using motion capture and 3D modelling techniques.*

Rotoscoping is another animation technique in which animators trace over live-action film movement, frame by frame, for use in animated films. It was originally used by animators to mimic real-life movements by pre-recording live-action film images that were then projected onto a frosted glass panel and re-drawn. The rotoscope projection equipment has now been replaced by computers and is referred to as interpolated rotoscoping. For example, interpolated rotoscoping was used to great effect in the 2006 film 'A Scanner Darkly'. To give the film its distinct look, the movie was filmed digitally and then animated using interpolated rotoscope over the original footage.

LINKS TO:

Unit 2: Design and Technology in Practice: Modelling and prototyping

THINK ABOUT THIS!

With more and more films using digital special effects, have directors come to rely on them to sell films? Is CGI over-used when traditional live action sequences would be as effective? Or do we expect to be amazed by 'out of this world' imagery?

Table 3.7 *Advantages and disadvantages of CGI.*

Advantages	Disadvantages
• The director can choose any angle for a scene, including angles that would have been hard or impossible in a live action movie. • Eliminates need for excessive and time-consuming green screen filming and set construction. • Limitless possibilities for rotating effect. • Costumes, make-up, body size and age can be changed to whatever is needed. • Characters blend in seamlessly with their digital environments. • Light, colours and filters are added digitally after filming.	• Expensive hardware and special programs are required to obtain and process the data. • Highly skilled technical animators required. • The technology becomes obsolete rapidly as better software and techniques are developed. • Movement that does not follow the laws of physics generally cannot be represented.

3 Biotechnology

Genetic engineering in relation to wood for the paper and board industry

The paper and board industry is placing greater demands on existing plantations and forests producing wood pulp because of the rapid growth in paper products. Only 10 per cent of land in the UK is forested: as a consequence, timber products, including wood pulp and paper goods, represent the UK's fourth-biggest import.

Biotechnology is at the forefront of research, experimentation and field trials into genetically modified (GM) trees for producing high yields of wood pulp. GM trees are the result of gene manipulation. This involves artificially inserting a gene from one plant into another, producing a change in the tree's biological characteristics. The main advantages to the paper and board industry from GM trees are:

- quicker-growing trees to provide a sustainable supply of wood pulp

- resistance to disease and insect attack to provide high-quality products

- reduced strength of lignin fibres enabling a reduction in the amount of chemicals needed in the paper-making process.

GM tree technology is gathering pace with field trials rapidly increasing around the world. Overall, genetic modification activities in forestry are taking place in at least 35 countries world-wide and are likely to make their commercial debut in Chile, China and Indonesia. There have been several trials in the UK so far, including GM Elm trees resistant to Dutch Elm Disease, which were successfully grown in Scotland.

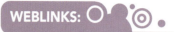

WEBLINKS:

www.ppic.org.uk – confederation of paper industries

THINK ABOUT THIS!

Scientists are artificially manufacturing new crops, materials and animals. How does this make you feel? The public reaction to GM crops and livestock has been quite negative – but what about the advantages? Think about the benefits to industry and its potential to solve problems in the developing world.

Biodegradable polymers

Biodegradable polymers are materials derived from renewable raw materials that will decompose in the natural environment. Biodegradation of polymers is achieved by enabling micro-organisms in the environment to break down the molecular structure of the polymer to produce an inert material that is less harmful to the environment. Many biodegradable polymers, such as polyhydroxyalkanoate (PHA), are fully biodegradable as they are derived purely from renewable sources. Other types are semi-biodegradable, mixing renewable sources with existing petroleum-derived synthetic polymers. At present, the

Table 3.8 Advantages and disadvantages of genetically modified (GM) trees in the production of paper and board.

Advantages	Disadvantages
<ul><li>Aids resistance of trees to disease.</li><li>Reduction of lignin in tree growth.</li><li>Reduces the toxic chemicals used in the paper industry needed to break down lignin.</li><li>Produces trees with increased growth rate.</li><li>Better forest management, which reduces deforestation.</li><li>Trees grown specifically for the paper industry.</li><li>Enzymes break down timber fibres more effectively.</li><li>Paper fibres can be more effectively bonded.</li><li>Recycled paper can be treated more effectively/easily.</li><li>Paper treated to biodegrade more easily and quicker.</li><li>Efficient and faster production.</li></ul>	<ul><li>Long-term side effects not yet apparent.</li><li>'Escape' of modified genes into natural ecosystems.</li><li>Development of tolerance to the modified trait by insects or disease organisms.</li><li>Rapid growth could cause shorter, more intensive rotations, resulting in greater water demand and reduced opportunity for nutrient recycling.</li></ul>

Table 3.9 Advantages, disadvantages and applications of biodegradable polymers.

Advantages	Disadvantages	Applications
• Fully degradable in suitable conditions, e.g. sun, moisture and oxygen. • Reduction of time in landfill and the associated harmful effects. • Starch-based plastics are formed from carbon that is already in the eco-system so does not contribute to global warming.	• Degradation of some plastics still contributes to global warming through the release of carbon dioxide as a main end-product. • Damages recycled plastics when mixed and reduces their value. • Fully biodegradable polymers are more expensive as they are not widely produced to achieve large economies of scale. • May not be as energy effeicent to produce as synthetic polymers, e.g. polypropylene. • Semi-biodegradable polymers remain in the environment for years.	• Packaging, e.g. blow-moulded bottles. • Disposable products used in the food industry, e.g. utensils and dishes. • Plastic wrap for packaging, e.g. moisture barrier films for hygienic products. • Coatings for paper and board. • Agricultural uses, e.g. slow-release pesticides and fertilizers, mulches that degrade over time. • Medical uses, e.g. gauzes, sutures, implants. • Pharmaceutical uses, e.g. coatings for pills. • New natural fibres for the textiles industry. • Replacement for expanded polystyrene.

use of biodegradable polymers is in its infancy. Once their production increases they can become more economically viable in relation to synthetic polymers and therefore provide a widespread subsitute.

Biopol®

Biopol is a trade name of the British chemical company ICI for the first fully biodegradable polymer, polyhydroxybutrate (PHB), developed commercially in the early 1990s. The first uses for Biopol were in the packaging industry to produce blow-moulded shampoo bottles as the material is water resistant and provides an effective barrier to air. 'Green' credit cards made from Biopol have been introduced as a replacement for some of the 20 million credit cards in circulation today.

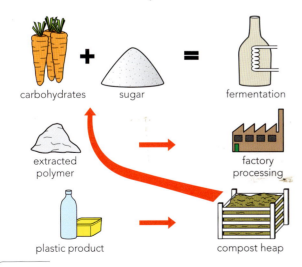

carbohydrates sugar fermentation

extracted polymer

factory processing

plastic product compost heap

Figure 3.11 The production of Biopol®.

Biopol is produced in nature through the fermentation by *Alcaligenes eutrophus* bacteria of sugars (glucose) and other carbohydrates that are collected in their cells as reserve material. Once this reserve material is separated and refined from the bacteria, a white powdered polymer is extracted. This polymer can then be used in the usual manufacturing processes to produce plastic products. The significance of this is that Biopol is produced naturally by renewable agricultural resources and, most importantly, it is fully biodegradable.

Biopol is stable when stored in air and is quite stable when stored even in humid conditions. Degradation to carbon dioxide and water will occur only when the polymer is exposed to micro-organisms found naturally in soil, sewage, river bottoms, and other similar environments. The rate of degradation is dependent on the material thickness and the number of bacteria present. Landfill simulations over a 19-week period show test bottles experienced a weight loss ranging from 30 per cent with oxygen present to 80 per cent with no oxygen present. The fact that Biopol decomposes more rapidly without oxygen present is significant because oxygen is not present in modern sealed landfills.

Innocent™ eco-bottle

Innocent™ has introduced an eco-bottle made from a material called polylactic acid (PLA), which is derived from corn starch. The use of PLA over regular PET and HDPE bottles has certain advantages.

• Regular plastic bottles use finite resources (e.g. oil and gas) whereas corn is a renewable resource.

- PLA is made using a totally carbon-neutral process – no greenhouse gases are emitted in its production.

- PLA is biodegradable, so it breaks down safely and relatively quickly.

Innocent™ is encouraging its customers to compost these bottles with all of their other green waste (e.g. garden waste, food scraps): the eco-bottle can be recycled along with regular plastic bottles in the normal way but, to maintain the quality of the recycled plastic, compostable bottles should not make up more than 1% of the mix. Innocent™ admits that these corn starch bottles are a very new development so there is currently a lack of composting infrastructure around. Only 10% of the UK's recycling centres have machines that can sort these bottles from the normal PET and HDPE plastic bottles. Therefore, the company has to be careful about how many compostable bottles it releases into the market before the correct facilities are in place to handle them.

WEBLINKS:

www.innocentdrinks.co.uk – eco-friendly drinks manufacturer

LINKS TO:

The use of biotechnology is directly related to issues regarding *Sustainability* (Unit 3).

Figure 3.12 *Innocent's eco-bottle is made entirely from a 100 per cent renewable source: corn. It is manufactured in a carbon-neutral process and is totally compostable, helping to reduce landfill waste.*

Systems and control

Getting started!

We are in a period of rapid development of manufacturing technologies, with many manufacturers now using complex computer controlled systems to produce their products. We notice this in the shops with a wide range of products on sale, especially electronics, at prices that seem to be getting cheaper all the time. As modern designers we must understand these systems, together with the impact that they will they have upon employment and our lives.

1 Manufacturing systems

Advanced manufacturing technology

Advanced manufacturing technology (AMT) describes the significant impact of computers on manufacturing. Computer technology has revolutionised the way that products are designed and manufactured. At every stage of the manufacturing process, computer technology is used to ensure fast, efficient high-quality production. The key facets of advanced manufacturing technology are given below.

Quick response manufacturing

Quick response manufacturing (QRM) was developed to make companies more efficient and hence profitable. QRM requires the manufacturer to move from traditional batch production to 'flow' production. In essence, QRM turns the company into one that responds to actual consumer demand rather than planning for an expected demand that may or may not happen.

QRM involves several concepts, such as total quality management (TQM), just-in-time (JIT) and manufacturing cells, but its main aim is to increase the overall flexibility and responsiveness of the company. For example, by manufacturing in cells, production teams can be dedicated to specific product lines. These teams can be quickly and efficiently re-allocated if the requirements change. Therefore, a manufacturer has increased production flexibility and will be better equipped to meet changing demands. In this way, no excess products are manufactured, only those that are actually needed.

In the ideal QRM situation, the manufacturer would begin production as soon as an order is initiated. Suppliers deliver raw materials directly to the production line, the product is manufactured and the finished goods would flow directly to a waiting truck for delivery. Therefore, QRM is described as a pull process because the raw materials are pulled through the production process according to market demands .

Table 3.10 Advantages and disadvantages to manufacturers of using quick response manufacturing (QRM).

Advantages	Disadvantages
• Less money needed to run the factory because fewer raw materials and finished goods are stocked. • Better position to increase market share as quicker response times may attract new clients. • Increased turnover of stock as production systems are triggered by demand. • Smaller batches are often produced, resulting in lower storage costs. • Reducing the cost of quality by minimising waste and by giving more responsibility to production teams.	• Increased reliance on suppliers to react to demand and quickly accommodate production schedules. • Poor supply could result in a manufacturer's inability to meet customer requirements. • Large variations in demand could cause problems if the manufacturer can not react to the high production volume efficiently. • Managing and implementing the change required can be very difficult as QRM changes the roles and responsibilities of employees.

Concurrent manufacturing

In order to remain competitive and cope with increasing market pressure from mounting customer demands, manufacturers need to get to market first with products that customers want. Concurrent manufacturing is about all of the key people who work at each stage of the design and manufacturing process working together to make sure that changes to one part of the process will not require changes to be made at another. In this way, designs become 'right first time'. This reduces product development times and enables the earlier release of new products.

Concurrent manufacturing systems aim to eliminate the need for design changes and overcome production problems and product introduction delays. Such delays and problems add significant costs to a product which in turn makes it less competitive than desired.

Concurrent manufacturing systems bring together members from a wide range of disciplines such as design, manufacturing, project management, technical support, marketing and other specialist areas to form a multi-disciplinary team. By working together, differences are more easily dealt with early in design and production, making the design process quicker and more efficient as, ideally, no re-design is necessary and manufacture can start earlier.

One of the most important factors in any successful concurrent system is the effectiveness of the project team. Therefore, excellent communication is needed to be truly effective. Computer-based systems enable efficient communication between individual team members and integrated project teams for product development. For manufacturing companies operating at a national or international level, a computer network is essential to support data transfer between team members. Concurrent manufacturing systems, effective management and teamwork ensure the development of a high-quality product at the right price in the shortest possible time.

Early in the development stage, designers can use quality function deployment (QFD) in order to create a more successful product. QFD is a method that incorporates customer satisfaction into the development of a product before it is manufactured. The main features of QFD are its focus on customer requirements, the use of multi-disciplinary teamwork and a comprehensive "House of Quality" matrix. This matrix is used by the team to translate customer requirements into a number of targets or specifications to be met by the new product. Some of the advantages of using QFD as part of a concurrent manufacturing system are:

- reduced time to market
- reduction in design modifications
- decreased design and manufacturing costs
- improved product quality
- enhanced customer satisfaction.

Figure 3.13 *Stages in concurrent manufacturing overlap significantly using multi-disciplinary design teams.*

Traditional sequential product development process:

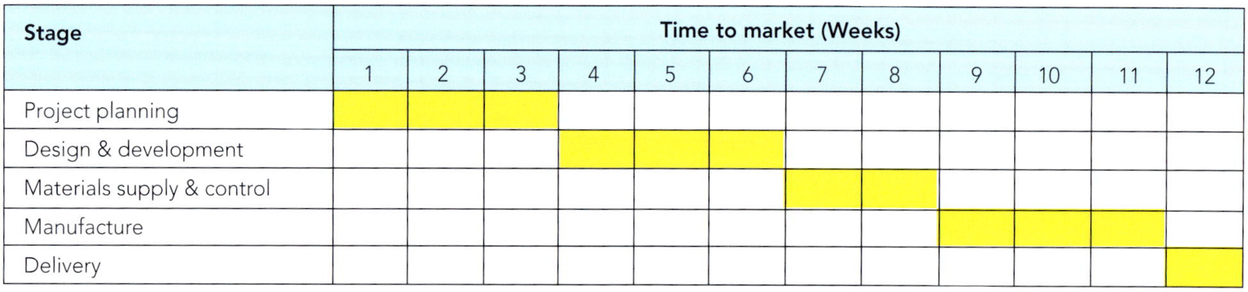

Stage	Time to market (Weeks)											
	1	2	3	4	5	6	7	8	9	10	11	12
Project planning	▮	▮	▮									
Design & development				▮	▮	▮						
Materials supply & control							▮	▮				
Manufacture									▮	▮	▮	
Delivery												▮

Development process using a concurrent manufacturing system:

Stage	Time to market (Weeks)											
	1	2	3	4	5	6	7	8	9	10	11	12
Project planning	▮	▮	▮									
Design & development		▮	▮	▮								
Materials supply & control			▮	▮								
Manufacture				▮	▮	▮						
Delivery						▮						

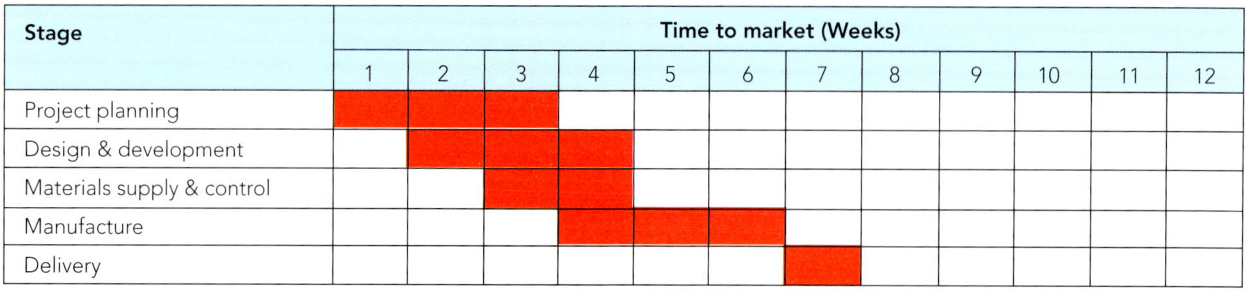

FACTFILE:

The six major components of the "House of Quality" matrix

	Component	Content
1	Customer requirements	A structured list of requirements taken from customer statements in a market survey.
2	Technical requirements	A technical specification is produced.
3	Planning matrix	A plan of what the product needs in order to meet customer and technical demands.
4	Interrelationship matrix	A grid showing how technical requirements dovetail with customer needs.
5	Technical correlation	An analysis of how the various technical requirements support or impede each other. For example, a high-quality finish such as 300 DPI and the need for low cost paper would demand a compromise or possibly a change to the required printing process.
6	Technical priorities, benchmarks and targets	Quality targets against each technical specification point can be measured. For example, the colour density on a print run.

Flexible manufacturing systems

A flexible manufacturing system (FMS) is one where several machines are linked together by a material-handling system such as a computer-controlled robot or conveyor system. An FMS brings together new manufacturing technologies such as CNC or robotics to form an integrated system. It is different from an automated production line because of its ability to process more than one type of product at the same time.

Modern FMS have powerful computing capacities that give them the ability not only to control and co-ordinate the individual equipment, but also to perform production planning. The main advantage of an FMS is its high flexibility in managing manufacturing resources like time and effort in order to manufacture a new product. This flexibility allows the system to react quickly to changes in production, utilising two main features:

- **Machine flexibility** – involves the system's ability to be changed to produce new product types, and the ability to change the order of operations carried out.

- **Routing flexibility** – involves the ability to use several machines at the same time to perform the same operation on a part thus increasing the speed of production. Also, these systems can readily adapt to changes in the product such as volume or size.

FMS vary in their complexity and size. Some are designed to be very flexible and to produce a large number of different parts in very small batches. Others have the ability to produce a single complete product in large batches from a sequence of many individual operations.

The advantages of flexible manufacturing systems are:

- increased productivity due to automation
- shorter lead times (the time from design to market) for new products
- lower labour costs due to automation
- improved production quality due to elimination of human error.

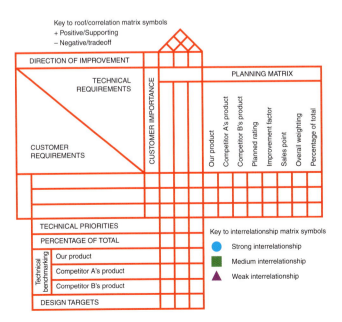

Figure 3.14 *An example of a 'House of Quality' matrix for assuring the quality of a design.*

2 Computer-integrated manufacture

Computer-integrated manufacture (CIM) takes the concept of integration of separate manufacturing technologies developed by FMS a step further by bringing together all aspects of a company's operations, not just those that are directly involved in manufacture. Under a CIM system all teams can share the same information and easily communicate with one another. A CIM system uses computer networks to integrate the processing of production and business information with manufacturing operations to create co-operative and smooth-running production lines. The tasks performed within CIM include:

- design of the product using CAD

- planning the most cost-effective workflow

- controlling the operations of machines and equipment needed to manufacture the product

- performing business functions such as ordering stock and materials and invoicing customers.

One of the drawbacks with a CIM system is its dependence upon computer data to fully integrate all operations. For example, the software from one brand of equipment may be incompatible or cause difficulties when integrated with

CNC machinery and an automatic storage and retrieval system (ASRS) from another brand. The cost of managing data is also a key issue within CIM. This is because if data becomes corrupted, it may cause machinery to malfunction. To prevent this, companies often use a product data management (PDM) system.

Product data management systems within CIM

PDM is an information system used to manage the data for a product as it passes from design to manufacture. The data includes plans, 3D models, CAD drawings, CNC programs and all related project data and documents. A PDM also highlights when changes are made to one database so that any effects on other parts can be determined.

By helping companies to manage data, PDM systems support efficient production. For example, the design of a component may go through many changes during the course of its development, each involving modifications to the CAD data. Once the designer is satisfied with the component, the PDM system may notify an analyst that the design is ready to perform a stress analysis on it. When that task is complete, the stress analyst performs an electronic sign-off. The PDM system will then notify a manufacturing engineer that the component is ready for planning of its manufacture and a tool designer that the component

Figure 3.15 *Product data management (PDM) systems enable efficient collaboration when planning production.*

design is ready for a tool design. After these tasks are performed, the various team members submit their data to the PDM system for a final review and team sign-off, after which it is released to begin manufacture.

The advantages of a product data management (PDM) system include:

- **reduced time-to-market** as data is instantly available to all teams for review, eliminating 'bottlenecks' where documents await distribution or sign-off

- **improved productivity** as changes to the product data are tracked and managed automatically, reducing the time taken to search and retrieve documents and giving the ability to re-use design data without repeating work

- **improved control** due to efficient management systems that assure everyone is working from the most current data.

Enterprise resource planning systems within CIM

Enterprise resource planning (ERP) systems attempt to combine all the software and data from various departments into one system that all can use. ERP, for example, improves the way in which a company takes a customer order and processes it into an invoice. ERP takes a customer order and provides a software 'road map' for automating the different steps along the path to fulfilling it. For example, when a customer order is entered into an ERP system, all the information necessary to complete the order is instantly accessible, such as the customer's credit rating and order history from the finance department, stock levels from the warehouse department and the delivery schedule from the logistics department. Employees in the different departments all see the same information and can update

it instantly. When one department finishes with the order it is automatically sent via the ERP system to the next department. Therefore, any order can be easily tracked and customers should receive delivery of their orders faster and without errors.

ERP systems are extremely expensive to install and costs are incurred during the 'switching over' period as a result of hardware investment and staff training. Ultimately, success depends on the skill and experience of the workforce and and the quality of ongoing training.

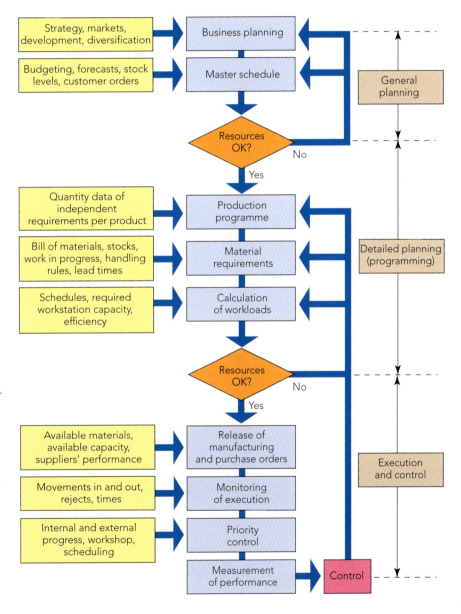

Figure 3.16 *Enterprise resource planning (ERP) in a modern computer-integrated manufacturing (CIM) system.*

Lean manufacturing and just-in-time systems

Lean manufacturing is, as the name suggests, manufacturing where there is no 'fat'. A key feature of lean manufacturing is the notion of 'just in time' (JIT), that is, there are no warehouses full of materials waiting to be used; materials arrive just when they are needed. The objective of lean manufacturing, then, is to provide techniques that ensure minimum waste is incurred during production and to produce products only when they are needed.

JIT is derived from a Japanese manufacturing philosophy. Quite simply, JIT ensures that the right materials, components and products arrive at the right time, at the right place, and in the exact amount. This reduces waste and overstocking as new stock is only ordered when it is needed, so saving warehouse space and storage costs. This focus on producing the right amount at the right time relies upon accurate analysis and forecasts using the right information at the right time. However, if a manufacturer is inaccurate in their predictions, such as a rise in demand for the product, stock will be used up rapidly with very little opportunity for re-supply. Production can also be held up or shut down completely if the raw materials supplier has problems fulfilling orders.

Table 3.11 The five key stages of lean manufacturing.

Key stage		Content
1	Value	Focus on value in the context of what the customer/end-user is prepared to pay for.
2	Value stream	Identify how value-adding and non-value-adding activities affect efficiency throughout production including: • for value-adding activities, machining, processing, painting, assembling, etc. • for non-value-adding activities, scrapping, sorting, storing, counting, moving, etc.
3	Flow	Design processes that result in uninterrupted flow, from raw materials to delivery of finished product.
4	Pull	Design manufacturing for 'pull' of product through the process as a response to demand, rather than 'push' of raw materials into the process by producing components irrespective of whether they are needed or not.
5	Perfection (Kaizen)	Adopt an approach that continually improves working processes .

Pull tools – Kanban

A lean organisation needs to be set up so that orders are pulled through the production system. In order to achieve this, companies have developed techniques to help them. One such system, known as Kanban, uses cards or containers as simple visual signals to indicate when to pull materials, components or products through the production system. The system relies on a simple rule of only producing or delivering when a card or empty container is passed to a workstation or manufacturing cell. By and large, production Kanban and transportation Kanban are combined into one integrated system. The production Kanban includes details of the operations that need to be carried out at the workstation or manufacturing cell, whereas the transportation Kanban only contains details regarding where the materials, components or products have come from and where they are going to.

The main benefit of using Kanban is that it reduces the amount of work-in-progress and finished goods in stock. Kanban restricts the supply of materials and components until they are needed, which provides an effective JIT system.

Perfection tools – Kaizen

Kaizen is also known as continuous improvement, where small changes are made to the production process resulting in small improvements being made. This system tends to be carried out on a regular basis as the changes made are often low cost and the improvements made tend to be small.

Flexible manufacturing cells

As customers demand variety and customisation of products as well as specific quantities delivered at specific times, a lean manufacturer must remain flexible enough to serve its customers' needs. Manufacturing cells allow manufacturers to provide their customers with the right product at the right time. They achieve this by grouping similar products into families that can be processed on the same equipment in the same sequence.

A manufacturing cell is a group of workstations, machines or equipment arranged such that a product can be processed progressively from one workstation to another without having to wait for a batch to be completed and without additional handling between operations.

Types of manufacturing cells

Functional cells	These cells perform a specific function as opposed to manufacturing a complete product and consist of similar equipment. For example, a factory that primarily carries out machining operations may have a bank of lathes together in a 'turning cell'.
Group technology (or mixed model) cells	These cells perform a series of operations for several different product lines. These products often involve very similar manufacturing operations though not usually identical. This type of cell can work very well within a lean manufacturing environment, particularly if the company is characterised by a large product range with low volume.
Product focused cells	These cells are product focused and typically manufacture one type of product through a series of operations. These are the ideal lean manufacturing cell and are perfect for a small product range with high volume manufacturers.

Cell production methods provide flexibility as upgrades to processes can be performed relatively quickly and easily by shutting down one cell whilst another simultaneously opens and takes its place. This means that the entire production line does not have to be disrupted, which reduces costs from stoppages, or 'downtime'. Flexibility also occurs when a large variety of products are assembled from a range of similar components. For example, cellular manufacture enables a large variety of cars to be produced using various combinations of customer options such as engine sizes, interior finishes, etc.

Typically, a manufacturing cell involves three to 12 workers and five to 15 workstations in a compact arrangement. However, some manufacturing cells are fully automated using CAM containing several CNC machines, computer-aided quality control and automated materials-handling systems. This is an ideal cell layout as it manufactures a narrow range of very similar products and is self-contained with all necessary equipment and resources. Materials sit in an initial queue when they enter the cell. Once processing begins, they move directly from process to process, resulting in a very fast throughput.

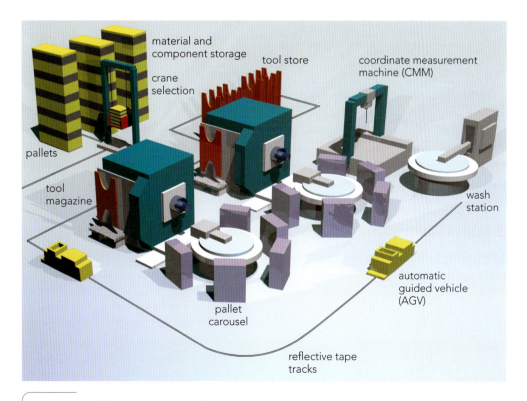

Figure 3.17 A fully automated manufacturing cell.

Table 3.12 Advantages and disadvantages of computer-aided manufacture (CAM).

Advantages	Disadvantages
• Greater control of the production process enabling fully automated production. • Safer working environments due to removal of risks to humans. • Flexible production as CNC machinery can be quickly re-programmed to manufacture alternative components, reducing set-up times. • Scale of production can be directly linked to customer demand, responding quickly to changes in quantities. • Improved productivity as production rates are consistent, less waste is generated and production costs are reduced. • Reduced manufacturing times as efficient cutting paths are generated by software. • Increased reliability and consistency in repetitive tasks.	• Extremely high set-up costs, as expensive machinery and installation are required. • Negative effects on employment as CAM requires less human involvement in its operation. • Worker morale may be affected due to 'machine minding' job roles.

Computer-aided quality control systems

Computer-aided quality control (CAQ) can be achieved within a manufacturing cell using a coordinate-measuring machine (CMM). A CMM is a mechanical system designed to move a measuring probe to determine the coordinates of points on the surface of a workpiece to accurately measure the sizes and positions of features on mechanical parts. This provides data that can immediately be fed back into the production process to analyse extremely small tolerances and control the quality of components. Laser scanning technology is advancing rapidly. Many thousands of points can be taken and used to not only check size and position, but to create a 3D image of the part. In addition, other systems can provide automatic identification such as optical character recognition and barcode readers. For example, in the food industry, a high-speed product identification system is designed to read a special code on every can of food at 1200 cans per minute. This prevents accidental product mixing and wrongly labelled products reaching customers.

Automated materials-handling systems

A materials handling system is a deveice that takes material from one place to another, such as a conveyor belt or truck. Within a factory, materials-handling starts with the unloading of goods from delivery transportation. The goods then pass through storage, onto machining, assembly, testing, packing and finally loading back onto transport for delivery. Each of these stages of the production process requires a slightly different design of handling equipment. For example, unloading from a lorry may require the manual operation of a fork-lift truck.

However, CIM systems enable a range of automated materials-handling systems to operate within the workplace. Fully automated handling systems ensure that the materials, components and assemblies are delivered to the production line when required without significant manual intervention.

Automatic storage and retrieval systems

An automatic storage and retrieval system (ASRS) is an automated robotic system for sorting, storing and retrieving items in a warehouse. Within CIM, the transportation of materials and components to the required points is controlled by computer. All stocks of materials and components are stored in racking systems. The ASRS system will select the correct component from the rack, retrieve it by means of a crane and place it on a conveyor or onto an automatic guided vehicle (AGV) for transportation.

Once completed, products are taken to a warehouse and from there they are distributed. Warehousing and distribution are about getting as many products as possible to market, in the shortest amount of time. While there are a variety of methods available, it is becoming increasingly necessary for companies of all sizes to replace manual methods with automated storage and retrieval systems. Traditional manual systems are up to four times slower than automated systems. In addition to increasing the efficiency of storage and retrieval and reducing labour costs, employees' technical skills are developed in the operation of a system.

Figure 3.18 *An automatic storage and retrieval system (ASRS) used in a warehouse.*

Automated guided vehicles

An AGV is a materials-handling device that is used to move parts between machines or work-centres. They are small, independently powered vehicles that are usually guided by radio frequency wires buried in the floor or use optical sensors in a laser-guided navigation system. They are controlled by receiving instructions either from a central computer or from their own on-board computer.

WEBLINKS:

www.amhsa.co.uk/case_studies.htm – Case studies from the Automated Material Handling Systems Association

Table 3.13 *Types and applications of automated guided vehicles (AGVs).*

AGV type	Application
Towing vehicles	These were the first type introduced and are still very popular today to pull a variety of trailer types.
Unit load vehicles	These are equipped with decks that permit unit load transportation and often automatic load transfer. The decks can either be lift and lower type, powered or non-powered roller, chain or belt decks or custom decks with multiple compartments.
Pallet trucks	These are designed to transport palletised loads to and from floor level, eliminating the need for fixed load stands.
Fork truck	These have the ability to service loads both at floor level and on stands. In some cases these vehicles can also stack loads in a racking system.
Light load	These are used to transport small parts, baskets, or other light loads through a light manufacturing environment and are designed to operate in areas with limited space.
Assembly line vehicles	These are an adaptation of the light load AGVs for applications involving serial assembly processes such as manufacturing cells.

The impact of advanced manufacturing technologies on employment

As with any computer-aided technology such as CAM, there has been some reduction of workforces as machines have become increasingly efficient. However, it does not eliminate the need for skilled professionals such as creative product designers, manufacturing engineers and CNC machine programmers. Computers, in fact, increase the levels of skill in many manufacturing professionals who use advanced productivity, visualisation and simulation tools.

The Manufacturing Institute in the UK argues that the negative aspects of manufacturing are magnified by the media. It outlines the 'top ten manufacturing myths' in order to promote manufacturing in a more positive light.

"Myth 01: Manufacturing jobs are all monotonous, strenuous and low paid"

"To the majority, the image most vividly conjured up when thinking of manufacturing is still one of endless assembly lines; employing poorly paid manual workers who carry out the same mundane tasks, whilst working in the same metre space, year upon year."

However, the Manufacturing Institute argues that:

"Over the past 20 years production lines have become increasingly automated, leading manufacturers to demand increased skills flexibility among their staff … Employers now require a multi-skilled workforce to service an increasingly challenging, diverse and multi-faceted industry … " Few industries have changed more than the print industry. The development of digital printing based around computers and laser technology has led to an increase in the need for a highly computer literate workforce.

"Myth 08: Manufacturing is not a creative industry"

"The case against this myth is simple – everything that is manufactured has to be designed … All components need to be functionally and creatively designed for purpose. A good functional design can save a company millions in production costs by eliminating waste, stock surplus and lead times, whilst a creatively designed unit will also help sell a product."

THINK ABOUT THIS!

Manufacturing industries are increasingly located in countries such as China rather than the UK. Why do you think this is? What could be done in the UK to promote careers in the manufacturing industries?

WEBLINKS:

www.manufacturinginstitute.co.uk – The Manufacturing Institute is an advisory service to manufacturers and universities in the North West

3 Robotics and artificial intelligence

Robots in automated manufacturing systems

The vast majority of robots in use today are found in the manufacturing industry on automated production and assembly lines and in manufacturing cells. Automation is the use of computer systems to control industrial machinery and processes, largely replacing human operators. The British Robot Association defines an industrial robot as:

"A re-programmable device designed to both manipulate and transport parts, tools, or specialised manufacturing implements through variable programmed motions for the performance of specific manufacturing tasks."

Japan is the world leader in robotics technology and they widen this definition to include arms controlled directly by humans which have a wide range of possible future applications.

The basic robotics technology in modern industrial robots is similar to CNC technology but most robots have many degrees of freedom. In manufacturing applications, robots can be used for assembly work, processes such as painting, welding, etc. and for materials handling. More recently robots have been equipped with sensory feedback through vision and tactile sensors. In the future robots may also link more intelligently with humans so that they can judge for themselves when it is safe to operate without having built-in guards and safety mechanisms that limit their operations.

WEBLINKS:

www.bara.org.uk/info_casestudies.htm – Case studies from The British Automation and Robot Association

FACTFILE:

Levels of complexity of manufacturing tasks for robots

Level	Applications
1	Applications that can be achieved using a simple robot using jigs and fixtures to position components and tooling to achieve the required accuracy. For example, spot welding, adhesive or sealant application and painting.
2	Applications requiring sensory feedback in order for small modifications to be made to the program to account for variation in the components. For example, arc welding, automotive window glazing and spare wheel mounting.
3	These applications require more complex sensory capabilities such as pattern recognition, that require complex decision making based on this feedback. For example, automated assembly processes such as locating and fixing wheels on a car.
4	The most difficult applications are those involving unpredictable behaviour of either the components or other equipment within the manufacturing cell. For example, operations such as handling of flexible components.

Increase in difficulty →

Table 3.14 *Advantages and disadvantages of robots for manufacturing.*

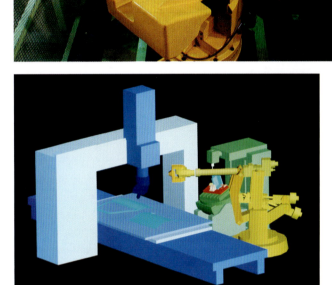

Figure 3.19 *Six degrees of freedom on a robotic arm make it flexible enough to carry out repetitive and precise tasks such as welding.*

Advantages	Disadvantages
• Ideal for repetitive, monotonous, mundane tasks requiring extreme precision.	• Robots do not have as impressive an array of sensors as humans (touch, vision, hearing, pattern recognition).
• Can be used in hazardous environments not suitable for human operators.	• Robots do not have the ability to learn and make decisions when the required data does not exist.
• Able to carry extremely heavy loads.	• Robots are not as flexible as humans and are harder to program to perform specific tasks.
• Highly flexible when responding to change as they are re-programmable.	• Robotics technology is extremely expensive to purchase and install in automated manufacturing.
• Can be programmed once and then repeat the exact same task for years.	• Human operators have to be excluded from robot working areas due to safety issues.
• Do not tire or suffer from lack of concentration and stress during repetitive tasks over long periods.	• High cost of making robot cells safe, including collision sensors.
• Cost effective as robots can operate continuously resulting in increased productivity.	• Maintenance issues as different brands of robots use different control systems, so maintenance crews need different specialist training.
• Produce high-repeatability, high-quality products using highly accurate inspection and measurement sensors.	• No standard robot programming language implemented, which can cause operating problems between different brands.

Industrial applications of artificial intelligence

A machine with artificial intelligence (AI) is one that exhibits human intelligence and behaviour and can demonstrate the ability to learn and adapt through experience. The main question is: can a machine really behave like a person? Research undertaken throughout the 1980s focused upon creating super-computers that could solve problems using reasoning skills like humans. However, humans have a consciousness that gives us feelings and makes us aware of our own existence, and scientists have found it extremely difficult to get robots to carry out simple cognitive (brain/thinking) tasks. Creating a self-aware robot with real feelings is a significant challenge faced by scientists hoping to mimic human intelligence in a machine. For this reason, development since the early 1990s has concentrated on developing expert systems and smaller, autonomous robots that mimic insect behaviour.

Expert systems

Expert systems are so called because of their ability to process large amounts of information and seemingly make decisions which are 'similar' to humans. In fact they are based upon logic systems. A very good example of an expert system is a computer controlled chess game which constantly processes the options available before making a move. In this way it appears to be operating like a human. In future, voice recognition systems coupled with expert systems could enable designers and manufacturers to talk with their computers in order to solve problems. For example, the system may ask the designer what help they need and automatically call up the appropriate information to help solve the problem. Expert systems could even function as co-workers, assisting and collaborating with design or operations teams for complex systems.

Autonomous robots

In manufacturing, the goal is to enable robots to learn the skills needed for any particular environment rather than programming them for a specific repetitive task. Intelligent machine vision systems are key to this as cameras are not as good as human optics. Humans can rely on guess-work and assumptions, whereas robots must comprehend an image by examining individual pixels, processing them and attempting to develop conclusions using expert systems. The advantage a robot may have over humans in this field is the ability to pick up infrared radiation (heat) as well as visible light for thermal imaging or even X-rays.

Figure 3.21 *The future of robotics – Honda's ASIMO can recognise moving objects, human postures, gestures and faces, and environments, and distinguish sounds.*

Autonomous robots in manufacturing must also be able to manipulate and 'feel' objects like humans. At present the typical robot arm allows six degrees of freedom to perform a small range of tasks and can be fitted with complex optical sensors. However, a human type hand is required with over twenty degrees of freedom fitted with hundreds of touch sensors with which the AI system can decide on the best way to manipulate an object from many possibilities. Teams of robots could then contribute to manufacturing by operating in a dynamic environment with minimal human interference.

4 Flow charts

A flow chart (see Figure 3.23) is a schematic representation of a sytem or process and indicates at which stage quality control should take place. It is a visual rather than text-based method of communication, which provides an easy reference for the viewer.

FACTFILE:

Flow chart symbols and their meanings

Symbol	Meaning
	START and **STOP** symbols are represented as lozenges (or rounded rectangles) and indicate the start or end of a process.
	PROCESSES or stages in production are represented as rectangles with the actual process written inside.
	DECISIONS are represented as a diamond (or rhombus) and can indicate where quality control should take place. This symbol has one arrow coming out from underneath for a YES decision and an arrow feeding back into the process stage indicating NO.
	ARROWS indicate the flow of control. An arrow coming from one symbol and ending at another symbol represents that control passes to the symbol the arrow points to.

Open-loop control systems

A system operating open-loop control has no feedback information on the quality of each stage of the process. Therefore the process will continue without any interference from the control system even when the output changes. This is a major disadvantage in an automated process, as it cannot detect or correct any errors in the process. An open-loop control system is often used in basic processes because of its simplicity and low cost, especially in systems where feedback is not critical. A light switch is an example of an open loop system – it is either on or it is off; there is no way of controlling the output.

Closed-loop control systems

Systems that utilise feedback are called closed-loop control systems. The feedback is used to make decisions about changes to the process, for example, QC. Closed-loop control systems have many advantages over open-loop controllers, including improved performance tracking throughout a production process and early detection of faults that can be rectified. By fitting a dimmer switch to the light switch above, the system now has feedback and the light can be dimmed to suit the user's needs. Most closed-loop systems are automatic, such as temperature controllers on air conditioning systems.

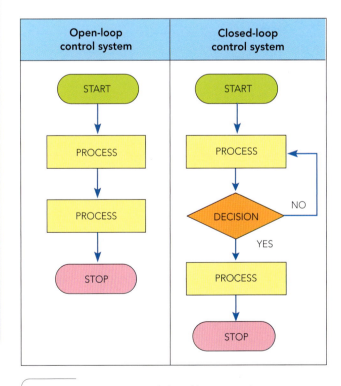

Figure 3.22 Basic open- and closed-loop control systems.

111

Text-based solution	Flow chart representation
• Machine is switched on. • Chamber needs to fill with plastic granules until it is totally full. • Plastic needs to be heated to correct temperature in order to melt. • Injection mechanism is operated. • Plastic needs to be injected into the mould until completely full. • Mould is split and moulded piece ejected. • Process is complete.	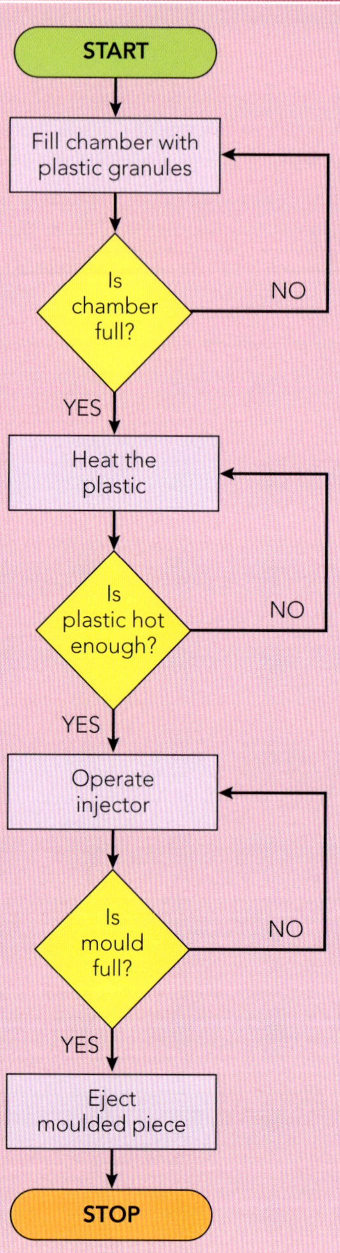 **START** → Fill chamber with plastic granules → **Is chamber full?** NO (loop back) / YES → **Heat the plastic** → **Is plastic hot enough?** NO (loop back) / YES → **Operate injector** → **Is mould full?** NO (loop back) / YES → **Eject moulded piece** → **STOP**

Figure 3.23 *Text-based solution represented as a flow chart for the injection moulding process.*

THINK ABOUT THIS!

All processes can be visualised as a flow chart. Why not try to construct flow charts of simple everyday processes to practice your skills? For example, stages in making a cup of tea including areas where decisions have to be made such as amount of sugar. Then move on to a more complex system such as using a vinyl cutter to produce an information sign.

Design in context

Getting started!

This section covers a wide range of design-related issues throughout history. Designers must be well informed, not only with current and future applications of technology, but with what has gone before. As a designer you cannot afford to design in a vacuum without external stimulus. Remember: all products were designed by someone for some purpose at some point in time – perhaps they could help with the design of your own!

1 The effects of technological changes on society

Design and technology have improved the lives of millions of people world-wide. But the changes brought about by developments in technology have resulted in far-reaching social consequences.

Mass production and the consumer society

The invention of the steam engine by James Watt in 1765 marked the beginning of the 'industrial revolution', which fundamentally changed life world-wide. Industrialisation and specialisation led to changes in production, the workforce, transportation and infrastructure. Many new fields of design were needed to accommodate this and the professional designer came into being. Population explosions occurred in towns and cities where production was centred and a new urban way of life was created. More people needed more products and mass production responded to this need. Expensive and time-consuming crafted work could now be replaced by machine work. Products once exclusively for the rich could now be made at an affordable price for ordinary working people.

Figure 3.24 *1950s youth culture saw the emergence of the 'teenager' as a mass consumer.*

The modern mass-consumer society is a feature of the affluent developed world where people's 'wants' are satisfied by a continual stream of new products. It is also referred to as a 'throw-away' culture with an increasing demand for convenience products such as fast food and over-packaged goods.

Mass consumerism, as we know it, developed during the 1930s out of popular culture, lifestyle and fashion. This was a time when international commerce and transportation systems developed and, with them, new opportunities for product design such as luxury ships, aeroplanes, hotels, theatres and department stores designed in an Art Deco style. Innovative new products and materials were introduced, especially in electrical consumer goods such as radios, refrigerators and washing machines. As people's standard of living improved, their demand for new products increased. Advertising and marketing became an important new industry, using market research, packaging and product styling to sell new products. The design of aesthetically pleasing products, most notably 'streamlining', became an important marketing tool.

After the Second World War (1939–1945), for many, there was a period of hardship with very few luxuries as countries struggled to recover from the fighting. By the mid-1950s, however, a new consumer society was developing. It started primarily in America and soon spread to Europe – the 'teenager' was born. Up until this point young men and women wore the same type of clothes as their parents and listened to the same type of music. The advent of Rock 'n' Roll was to change all that and soon teenagers were rebelling against their parents' values and began carving out a style all of their own. Design evolved rapidly to meet the expanding needs of the teenage market, incorporating high fashion and consumer goods such as portable transistor radios for the beach and cars, motorbikes and scooters for the increasingly mobile and independent youth culture.

LINKS TO:

Unit 3: Design in Context: Influences of design history on the development of products: Art Deco and Streamlining.

Targeting children as new consumers

A rather disturbing feature of modern mass consumerism is the targeting of young children by marketing companies in order to stimulate interest in products at an early age.

For example, a new blockbuster movie is released with a wide range of associated merchandise such as toys and lunchboxes that children will pester their parents for. In the past, marketing to children was restricted to toys and sweets and was relatively low-budget. However, as TVs and computers have made their way into children's bedrooms, they can now be targeted not only through TV adverts but the Internet and e-mail as well. It has even been documented that some marketing companies have posed as children in chat rooms to stimulate interest in certain products.

Sue Palmer, in her book *Detoxing Childhood*, outlines the key marketing strategies currently used to target children as new consumers.

- *'Kids are growing older younger'*, which exploits children's natural yearning to play at being grown-ups and targets them with mini-supermarkets and mini-briefcases that emulate the real thing.

- *'Gotta catch 'em all'*, which plays on a child's natural urge to collect things, such as Pokemon trading cards, with new sets being released regularly so it is difficult to actually collect them all.

- *'The culture of cool'*, which plays on a child's need to be accepted by wearing the right brands.

THINK ABOUT THIS!

What is your opinion on this modern marketing strategy to target young children? Do you think it is right to bombard children with advertising messages at such an early age? What kind of pressure does this put on children and adults?

Figure 3.25 *Movie merchandising targets children as new consumers.*

Built-in obsolescence

Built-in or planned obsolescence is a method of stimulating consumer demand by designing products that wear out or become outmoded after limited use. In the 1930s an enterprising engineer working for General Electric proposed increasing sales of flashlight lamps by increasing their efficiency and shortening their lifespan. Instead of lasting through three batteries he suggested that each lamp last only as long as one battery. By the 1950s built-in obsolescence had been routinely adopted by a range of industries, most notably in the American motor and domestic appliance sectors. Nowadays products such as laptop computers are obsolete as soon as they are purchased.

LINKS TO:

Unit 3: Sustainability: Cleaner design and technology.

THINK ABOUT THIS!

Make a list, under the headings in the table below, of products that you have consumed over the past 12 months. For example, how many different mobile phones or MP3 players have you owned and where are they now? Does this type of built-in obsolescence bother you?

Mass production and its effect upon employment

Mass production processes, as a result of the industrial revolution, meant that the craftsperson was replaced by low-skilled workers in highly mechanised factories. What started out as a wonderful opportunity for ordinary people to find work and gain access to inexpensive consumer products ended in misery for many. Low skills equalled low wages and the employment of women and children in 'sweatshop' type factories. The resulting poverty led to workers' uprisings and the development of trade unions aimed at combating poor living conditions, poverty and the increasing pollution brought about by industrialisation.

Although working conditions have generally improved, modern mass production still has some very negative social consequences. The use of highly automated production and assembly lines has reduced the workforce required in many factories. The resulting jobs can be divided into two main categories: high-skilled technical roles and low-skilled manual roles. Higher-paid technical roles are required to set up and maintain machinery. Low-skilled and often low-paid workers are utilised on production lines for specialist repetitive tasks, which can lead to very poor job satisfaction and morale.

The 'new' industrial age of high-technology production

In the 20th century, developments in materials and manufacturing technologies, together with changes in lifestyle, revolutionised product design. New materials

Table 3.15 Forms of built-in obsolescence.

Form of obsolescence	Description	Example
Technological	Occurs mainly in the computer and electronics industries where companies are forced to introduce new products with increased technological features as rapidly as possible to stay ahead of the competition.	Mobile phones with image capture almost immediately superseded by phones with moving-image capture.
Postponed	Occurs when companies launch a new product even though they have the technology to realise a better product at the time.	It is not unrealistic to imagine that when Sony launched its PS3 games console it knew what its next generation games console would look like – the PS4?
Physical	Occurs when the very design of a product determines its lifespan.	Disposable or consumable items such as light bulbs and ink cartridges for printers.
Style	Occurs due to changes in fashion and trends where products seem out of date and force the customer to replace them with current 'trendy' goods.	High-street Summer/Winter fashion collections. Premiership clubs update their kit every season so fans need to constantly purchase new replica football kits.

such as metal alloys, polymers and composites enabled new ways of designing and manufacturing. In particular, the development of digital computers in the 1940s and the silicon chip in the 1960s enabled relatively inexpensive portable computer technology, which transformed modern industrial society.

Computers in the development and manufacture of products

CIM systems incorporating CAD and CAM have revolutionised modern manufacturing and the print industry. The digital age has brought about change to which business has responded by providing quick-turnaround jobs to meet client needs. Printers are capable of producing short full-colour runs with extremely fast delivery times and product designers are able to drastically reduce development times and costs.

On-demand printing quickly supplies the exact amount of copies to satisfy each customer's needs. The use of computers in pre-press means that information can be stored and transferred digitally so designs can be quickly developed in consultation with the client. Once designs are finalised, printing plates can quickly be produced using computer-to-plate (CTP) technology. This cuts out the long process of producing printing plates and instead data is transferred directly to a laser engraver that forms the plate. Printing costs can be significantly reduced with digital printing machines that can operate up to 14,400 pages per hour. Digital printing is well suited to the production of short print runs as it does not require the making of printing plates, unlike commercial printing processes such as offset lithography. In post-press, the printed materials can be die-cut, folded, glued or bound using automated machinery with efficient workflows.

LINKS TO: ○ ◎ •

Unit 2: Industrial and commercial practice: Information and communication technology (ICT).

Unit 2: Systems and control: Manufacturing systems, CIM, robotics and artificial intelligence (AI).

Miniaturisation of products and components

The most important technological development in recent years has been in the field of microelectronics. Not only

have products reduced in size through technological advances but multi-functional products have become possible. For example, the mobile phone has reduced in size considerably from models first introduced in the 1980s, when most were too large to be carried in a jacket pocket so they were typically installed in vehicles as car phones. The miniaturisation of mobile phones has been possible due to three key developments.

- **Advanced integrated circuits** (ICs) or microprocessors that allow more circuitry to be included on each microchip, increasing functionality and power.

- **Advanced battery technology** including Lithium-Ion rechargeable batteries, providing a lightweight means of storing a lot of energy resulting in smaller and thinner fuel cells.

- **Advanced liquid crystal displays** (LCDs) enabling colour screens that are thinner and brighter and require much smaller current, meaning greater energy efficiency and slimmer housings.

The widespread use of these technologies has also led to advances in manufacturing that have reduced unit costs considerably, enabling low-cost electronic products.

The mobile phone is now much more than a telephone – it has become multi-functional. Communication, entertainment and computing services are converging within the same device, offering substantial choice to consumers. Mobile phones often have features beyond sending text messages and making voice calls. Product convergence has enabled Bluetooth connectivity, Internet access, built-in cameras and camcorders, games and MP3 players to be included on a single device.

Figure 3.26 *The miniaturisation of mobile phones has been possible through advances in technology.*

Use of smart materials and products for innovative applications

The continued development of smart materials has seen them being applied to a whole range of innovative products and systems where their ability to respond to changes and return to their original state is a real advantage.

LINKS TO:

Thermochromic liquid crystals, piezoelectric crystals and smart ink are explored in *Unit 2: Materials and components: smart materials*

Table 3.16 *The use of smart materials and products for innovative applications.*

Smart material	Application	Advantages	Disadvantages
Smart glass	Used to change light transmission properties of windows or skylights when a voltage is applied i.e. changes opacity from transparent to translucent.	• Controls amount of heat passing through a window, saving energy costs. • Provides shade from harmful UV rays. • Provides privacy.	• Expensive to install. • Requires constant supply of electricity. • Speed of control. • Degree of transparency.
Shape memory alloys (SMAs)	Used in spectacle frames as the crystal structure of this advanced composite, once deformed, can regain or 'remember' its original shape, e.g. Memoflex glasses.	• Superelasticity – extremely flexible so can be bent or 'sat on' without permanently deforming. • Immediately recovers original shape. • Lightweight and durable – alloy contains titanium.	• Not unbreakable. • More expensive than similar polymer frames.
Thermochromic pigments	Combined with polymers and used in 'chameleon' kettles, which change colour when boiling (bright pink) and return to original colour when cool (bright blue).	• Immediate visual indication of temperature. • Safety feature. • Aesthetic 'novelty' appeal.	• Limited colour range. • Not possible to engineer accurate temperature settings to colour changes.
Smart fluid/oils/grease	Used in a car's suspension system to dampen the ride depending upon road conditions, e.g. second-generation Audi TT. The fluid contains metallic elements that alter the viscosity of the fluid when a magnetic field is applied.	• Improves handling and road-holding as it adapts to road. • Better and faster control.	• More expensive than traditional systems.

Figure 3.27 *Audi's 'magnetic ride' system uses smart fluids to adjust the car's suspension.*

The global marketplace

The need to be competitive means that many companies sell their products all over the world. It can sometimes be a problem to design for unfamiliar markets or design products that will sell across different countries. Many companies employ design teams situated throughout the world so they can design for a particular local market or culture. Other companies use focused market research to discover the needs of specific markets.

Offshore manufacturing of multinationals

Offshore manufacture is a driving force in the global marketplace. There is an increased awareness by multinational companies based in developed countries (usually in the West and Australia) of the value of offshore manufacturing as a vital strategic tool. Many companies will draw upon the individual expertise of other countries to develop new products, especially in the field of technology.

Companies are relocating to less-developed countries such as India, China and former Soviet nations and outsourcing their work. Modern corporate buildings and industrial estates are sprouting up in these countries to supply the new demand for outsourcing and offshore manufacturing. Initially jobs in developing countries were created through the manufacture of shoes, cheap electronics and toys, and subsequently simple service work such as processing credit card receipts. Now all kinds of 'knowledge work' and manufacturing can be performed almost anywhere. For example, there is a trend for call centres dealing with the UK public to be based in India.

The driving forces are digitisation, the Internet and high-speed data networks that cover the entire globe. Design data can simply be sent to another country for manufacture or localised expertise can provide the design and development of products. Why do multinationals manufacture offshore, or outsource? The answer is quite simple: it costs them less. It is now possible to receive the same quality work at a fraction of the cost than if Western companies manufactured in their own countries. For example, mould-making for the purpose of injection moulding is generally much more affordable in China than in the West (about 50% lower in China, 30% lower

in Taiwan). In addition, by having bases in developing countries it is possible to gain greater access to expanding overseas markets.

Obviously this calls into question certain ethical issues such as large-scale unemployment in developed countries and exploitation of labour in developing countries. For instance, why would a British-based multinational company continue to pay the minimum wage to its UK employees when they could employ Indian or Chinese labour for 50–60% less? However, workers in developing countries may not be given the opportunities for promotion, pay rises, company benefits, union membership and working conditions that their Western colleagues demand as basic human rights. As multinationals build centres of operation and factories in these areas the local workforce is displaced from their traditional trades and become more dependent upon the largely unskilled labour that many industrial processes require.

THINK ABOUT THIS!

Discuss the effects, both positive and negative, of the use of offshore manufacturing and outsourcing in relation to:
a) multi-national companies
b) workers in developing countries
c) workers in developed countries.

Local and global production

Issues relating to local and global production are concerned with the effects of the global economy and of multinationals on quality of life, employment and the environment. Whilst the headquarters of multinationals are often located in developed countries, some multinationals are based in developing countries. Though economic regeneration is generally welcomed by the governments of developing countries, there are also a number of negative effects on the local population.

Table 3.17 *Advantages and disadvantages of global manufacturing in developing countries.*

Advantages	Disadvantages
• Economic regeneration of local area through increased employment in manufacturing and service industries. • Improvement in living standards through career development and multi-skilling of workforce. • Physical regeneration of local area through development of infrastructure, transportation and/or local amenities. • Widening of the country's economic base and enabling of foreign currency to be brought into the country, which improves their balance of payments. • Enabling of the transfer of technology that would be impossible without the financial backing of multinationals.	Environmental issues: • increased pollution and waste production as a result of manufacturing activities. • destruction of local environment to build factories, processing plants, infrastructure, etc. Employment issues: • lower wages than workers in developed countries where a minimum wage operates • promotion restrictions as managerial roles occupied by employees from developed countries • no unions for equal rights issues including unfair dismissal/hire and fire • lower safety standards when using 'sweat shops' • devaluing of traditional craft skills, replacement by repetitive 'machine minding' tasks • local community can become dependent on multinationals, leaving community devastated if the multinational pulls production.

2 Influences of design history on the development of products

All design has either been commissioned by or produced for the specific needs of somebody at some point in time. The history of design is inextricably linked to the social, political and economic history of the modern world. Designers have always looked for inspiration from other times and cultures and taken advantage of new technologies, which has had a major effect on the design of their products.

Arts and Crafts (1850–1900)

Philosophy

The Arts and Crafts movement grew out of a concern for the effects of industrialisation upon design, traditional craftsmanship and the lives of ordinary 'working class' people. Although the technical advances of the 19th century brought about new production processes, the design of mass-produced products, such as furniture, was often overlooked. Therefore, poor-quality, over-decorated and often oversized imitations of traditional items of furniture were being produced. This type of furniture was totally inappropriate for the majority of ordinary people

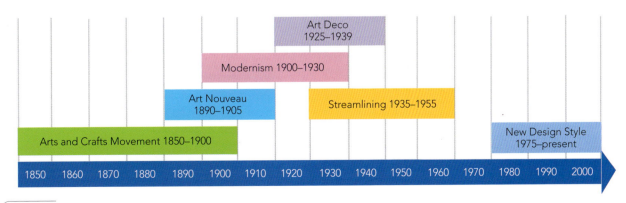

Figure 3.28 *This timeline demonstrates the overlapping of design movements and that one movement did not simply end and another take its place. (Note that the design period between 1955 and 1975 is not covered by the Edexcel specification.)*

who required simple and inexpensive products for their cramped living conditions.

Around this time emerged the two founding figures of Arts and Crafts: John Ruskin (theorist and critic) and William Morris (designer, writer and activist). Ruskin examined the relationship between art, society and labour. Morris put Ruskin's philosophies into practice, placing great value on craftsmanship, simple forms and patterns inspired by nature and the beauty of natural materials. In response to the effects of industrialisation, they helped establish a number of workers' guilds and societies to break down the barriers between architects, artists, designers and makers and pioneered new and unified approaches to design and decorative arts. Their ideas came from the conviction that traditional arts and crafts including weaving, carpentry and stained glass as a 'cottage' industry could change people's lives by empowering the individual as designer/maker of their own products.

Style

- **Simplicity** – Interiors were visually simplistic by removing clutter and including suitably proportioned furniture, which would provide a practical and clean living environment. Furniture was 'humbly' constructed with minimal ornate decoration. The roughness and simplicity of some work was shocking: one reviewer in 1899 referred to an Arts and Crafts piece as looking 'like the work of a savage'.

- **Splendour** – The arts and crafts approach to design led to designers often experimenting with different materials and new techniques in artistic ways. Therefore, small and highly ornate artefacts were produced working with unusual materials and precious metals.

- **Nature** – Natural plant, bird and animal forms were a powerful source of inspiration. The use of stylised flower patterns emulating the natural rhythms and patterns of plants and flowers were a reflection of a purity of approach. Symbolism with motifs such as the heart symbolising friendship or the sailing ship representing the journey of life into the unknown appear in many pieces of work.

- **Colour and texture** – Colour was used in Arts and Crafts interiors to provide unity and focus. The link between colour and nature was particularly close in Arts and Crafts style. Architects and designers preferred natural materials: stone, wood, wool and linen, and materials that were available locally. Rich materials, highly decorated surfaces and strong colours tended to be concentrated in small areas.

Figure 3.29 *Arts and crafts interiors were based around the simple rural way of life.*

William Morris (1834–1896)

"Have nothing in your houses that you do not know to be useful, or believe to be beautiful." (William Morris, *The Beauty of Life*, 1880.)

William Morris was a poet, writer, designer and innovator in the Arts and Crafts movement but, above all, he was a socialist.

At university, Morris and his friends were influenced by the writings of the art critic, John Ruskin, who praised the art of medieval craftsmen, sculptors and carvers whom he believed were free to express their creative individualism. Ruskin was also very critical of the artists of the 19th century, whom he accused of being "servants of the industrial age". After university they formed their own company of designers and decorators with the emphasis placed upon traditional craftsmanship and natural materials. Morris, Marshall, Faulkener & Co specialised in producing stained glass, carvings, furniture, wallpaper, carpets and tapestries. The company's designs brought about a complete revolution in public taste.

Figure 3.30 *The Red House, built in 1859 for William Morris, reflected his 'country cottage' idealism.*

Despite the large number of commissions that he received, Morris continued to find time to write poetry and prose and had a number of his works published. His passion for creating 'fantasy worlds' in his novels is said to have had a direct influence on J.R.R. Tolkein's *Lord of the Rings*.

In the 1870s Morris became upset by the aggressive foreign policy of the Conservative Prime Minister, Disraeli, and disillusioned with the subsequent Gladstone Liberal Government. In 1884, Morris co-formed the Socialist League. Strongly influenced by the ideas of Morris, the party published a manifesto where it advocated revolutionary international socialism. Over the next few years Morris wrote socialist pamphlets, sold socialist literature on street corners, went on speaking tours, encouraged and participated in strikes and took part in several political demonstrations (on which he was once arrested). These strong socialist beliefs directly influenced his design philosophy of simple, natural products produced by the individual rather than mass-produced by large-scale industry.

Figure 3.31 *The floral ornamentation of Morris's patterns drew heavily upon natural form*

After Morris's death in 1896 the business continued until 1940; a large textile company bought many of Morris's designs and printing blocks for fabric and wallpaper. Many William Morris prints are still in production and have influenced the design style of large companies such as Laura Ashley.

WEBLINKS:

www.artsandcraftsmuseum.org.uk
www.blackwell.org.uk
www.morrissociety.org

Art Nouveau (1890–1905)

Philosophy

Art Nouveau or 'new art' was an international style of decoration and architecture that developed in the late 19th century. The name derives from the *Maison de l'Art Nouveau*, an interior design gallery opened in Paris in 1896, but in fact the movement had different names throughout Europe.

It was developed by a new generation of artists and designers who sought to fashion an art form appropriate to their modern age. The underlying principle of Art Nouveau was the concept of a unity and harmony across the various fine arts and crafts media and the formulation of new aesthetic values. It was during this period that modern urban life, as we recognise it today, was also established. Old traditions and artistic styles sat alongside new, combining a wide range of contradictory images and ideas. Many contemporary artists, designers and architects were excited by new technologies and lifestyles, while others retreated into the past, embracing the spirit world, fantasy and myth.

Art Nouveau forms a bridge between the Arts and Crafts and Modernism. There was a strong link between the decorative and the modern that can be seen in the work of individual designers. Many Art Nouveau designers appreciated the benefits of mass production and other technological advances and embraced the aesthetic possibilities of new materials. In architecture, for example, glass and wrought iron were often creatively combined in preference to traditional stone and wood.

Figure 3.32 Paris Metro station by Hector Guimard, c.1900, creatively combines glass and 'organic' wrought ironwork.

Figure 3.33 *A luxurious Tiffany lamp, c.1900, demonstrates master craftsmanship.*

Others deplored the shoddiness of mass-produced machine-made goods and aimed to elevate the decorative arts to the level of fine art by applying the highest standards of craftsmanship and design to everyday objects.

Style

- **Nature** – Like the Arts and Crafts before them, Art Nouveau designers were heavily influenced by natural forms and interpreted these into sinuous, elongated, curvy 'whiplash' lines and stylised flowers, leaves, roots, buds and seedpods. As a complement to plant life, exotic insects and peacock feathers often featured in Art Nouveau designs.

- **The female form** – Art Nouveau is frequently referred to as 'feminine art' due to its frequent use of languid female figures in a pre-Raphaelite pose with long, flowing hair.

- **Other cultures** – The arts and artefacts of Japan were a crucial inspiration for Art Nouveau. Japanese woodcuts, with asymmetrical outlines and the minimal grid structures of Japanese interiors provided vertical lines and height. Celtic, Arabian and ancient Greek patterns provided inspiration for intertwined ribbon patterns.

Figure 3.34 *Mucha's poster for Job cigarette papers, 1898. Interest in Mucha's distinctive style experienced a strong revival in the 1960s and is particularly evident in the psychedelic artwork for many pop groups of this era.*

Charles Rennie Mackintosh (1868–1928)

In Britain the Art Nouveau style was exemplified by the work of Charles Rennie Mackintosh. Born in Glasgow, Mackintosh was interested in a career as an architect from an early age, and when he was 16 he became an apprentice to a Glasgow architect, studying at the same time as an evening student at the Glasgow School of Art. It was here that he met like-minded artists and formed the 'Glasgow Four', including his future wife Margaret Macdonald. Through their paintings, graphics, architecture, interior design, furniture, glass and metalwork they created the 'Glasgow style' of Art Nouveau, which influenced many designers throughout Europe.

In 1889, Mackintosh joined the firm of Honeyman & Keppie where he remained until 1913, becoming a partner in 1904.

Figure 3.35 *The Glasgow School of Art demonstrates Mackintosh's modern style, which was rooted in traditional Scottish architecture.*

Figure 3.36 *This Mackintosh interior demonstrates his trademark style of a crisp white linear finish dispersed with rose motifs.*

All his most important architectural and decorative work was achieved during this period. It is clear that he was allowed a degree of autonomy within the firm, developing his own markedly individual style in a way that is not usually possible for a man without his own independent practice. In 1896, Mackintosh won the competition for the building of the new School of Art in Glasgow – a project which gave him an international reputation.

It is perhaps Mackintosh's interior designs that best highlight his goal to create a new artistic harmony. The unification of architectural elements, furniture, furnishings and decoration produced highly aesthetic yet practical domestic and commercial environments. He designed all of the furniture, fixtures and fittings in all of his projects. His style incorporated a contrast between strong right angles and floral-inspired decorative motifs with subtle curves, along with some references to traditional Scottish

architecture. His designs for a 'House for an Art Lover' for an international competition in 1901 brought him great praise, although he was disqualified due to late entry. In tribute to his thoroughly modern style the house was built in the 1990s and can be visited by the general public.

Other notable domestic Mackintosh designs include Windyhill, 1900 and The Hill House, 1902. His experimentation with the possibilities of commercial production is best illustrated by The Willow Tea Rooms, 1903.

WEBLINKS:

www.crmsociety.com
www.charlesrenniemac.co.uk
www.houseforanartlover.co.uk

Modernism (1900–1930)

Modernist architects and designers rejected the old style of designing based upon natural form and materials. They believed in 'the machine aesthetic', which celebrated new technology, mechanised industry and modern materials that symbolised the new 21st century. Modernist designers typically rejected decorative motifs in design and the embellishment of surfaces with 'art', preferring to emphasise the materials used and pure geometrical forms.

Modernist principles soon spread throughout Europe with groups including De Stijl in the Netherlands, Bauhaus in Germany, Constructivism in Russia and Futurism in Italy. Le Corbusier, a french architect, thought that buildings should function as "machines for living in" where architecture should be treated like the mass-production of products. This resulted in many high-rise blocks of flats with repetitive 'cubes' as living spaces. Architect Ludwig Mies van der Rohe adopted the motto "less is more" to describe his minimilist aesthetic of flattening and emphasising the building's frame, eliminating interior walls and adopting open-plan living spaces.

The Bauhaus (1919–1933)

The German economy was in a state of collapse following Germany's defeat in the First World War. A new school of art and design was opened in Weimar to help rebuild the country and form a new social order. Walter Gropius was appointed to head the new institute and named it Bauhaus meaning 'house for building', which was to combine all the arts in ideal unity.

Philosophy

The central idea behind the teaching at the Bauhaus was a range of productive workshops where students were actively encouraged to be multi-disciplined and trained to work with industry. The Bauhaus contained a carpenters' workshop, a metal workshop, a pottery, and facilities for painting on glass, mural painting, weaving, printing, and wood and stone sculpting. Gropius saw the necessity to develop new teaching methods and was convinced that the base for any art was to be found in handcraft: "The school will gradually turn into a workshop." Indeed, artists and craftsmen directed classes and production together at the Bauhaus in Weimar. This was intended to remove any distinction between fine arts and applied arts. The Bauhaus workshops successfully produced prototypes for mass production: from a single lamp to a complete dwelling.

The Bauhaus school disbanded in 1933 when Adolf Hitler and the Nazi party rose to power in Germany before the start of the Second World War. Many Bauhaus leaders, including Gropius, emigrated to the United States to avoid persecution, where they continued to practice. The term 'International Style' was applied to this American form of Bauhaus architecture.

Style

- **'Form follows function'** – Bauhaus featured functional design as opposed to highly decorative design. Designers produced high-end functional products with artistic pretensions which primarily worked well but also looked good. Simple, geometrically pure forms were adopted with clean lines and the elimination of unnecessary clutter.

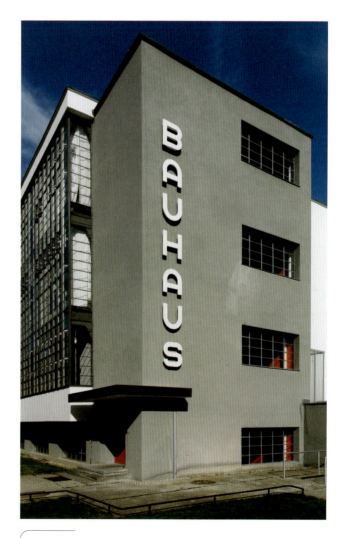

Figure 3.37 *The Bauhaus school, which educated students in art and design with industrial applications.*

- **'Products for a machine age'** – Products respected the use of modern materials such as tubular steel and mechanised mass production processes. As a result products looked like they had been made by machines and were not based upon natural forms as with previous movements.
- **'Everyday objects for everyday people'** – Consumer goods should be functional, cheap and easily mass-produced so that ordinary working-class people could afford them.

Marcel Breuer (1902–1981)

Marcel Breuer was born in Hungary and worked in an architect's office in Vienna before going to Weimar to study at the Bauhaus from 1920 to 1924. After his trade test he became the manager of the furniture workshop, stressing the combination of art and technology, and created his best-known piece called the 'Wassily' chair. It was here that he met the constructivist artist Wassily Kandinsky who also lectured at the school, but despite popular belief, the chair was not actually designed for Kandinsky.

The Wassily chair, also known as the Model B3 chair, was designed by Breuer in 1925–1926. Kandinsky had simply admired the completed design, and Breuer fabricated a duplicate prototype for Kandinsky's personal quarters. The chair became known as "Wassily" decades later, when it was re-released by an Italian manufacturer who had learned of the Kandinsky connection in the course of its research on the chair's origins.

Figure 3.38 Marcel Breur's iconic 'Wassily' chair demonstrates the principles of Bauhaus design.

The Wassily chair was revolutionary in the use of the materials (bent steel tubes and leather) and methods of manufacturing. The design was only technogically feasible because a German steel manufacturer had recently perfected a process for mass-producing seamless steel tubing. Previously, steel tubing had a welded seam that would collapse when the tubing was bent. The Wassily chair, like many other designs of the modernist movement, has been mass-produced since the 1960s, and as a design classic is still available today.

In 1937, Breuer emigrated to the United States and received a professorship at the School of Design at Harvard University. In 1946 he founded his own company in New York, Marcel Breuer & Associates, which he managed until his retirement in 1976.

WEBLINKS:

www.bauhaus.de/english
www.artsmia.org/modernism
www.reeform.com

THINK ABOUT THIS!

The principles of modernist architecture were applied to many high-rise blocks of flats built after the Second World War in order to give working-class people modern and affordable housing.

However, a large number of these blocks of flats have since been demolished. Discuss the reasons for failure of many works of modernist architecture with particular reference to the materials used in their construction and the effects of high-rise living upon inhabitants.

Art Deco (1925–1939)

Philosophy

The term Art Deco is widely used to describe the architectural and decorative arts style that emerged in France in the 1920s. It took its name from the 1925 Exposition des Arts Decoratifs held in Paris to celebrate the arrival of a new style in applied arts and architecture. It was an eclectic style that drew on tradition and yet simultaneously celebrated the mechanised, modern world.

became the popular face of modernism and its influence was witnessed world-wide.

Style

- **Geometric forms** – Popular themes in Art Deco were trapezoidal, zig-zagged, geometric fan motifs. Sunburst motifs, for example, were widely used in such varied contexts as ladies' shoes, radiator grilles, the auditorium of the Radio City Music Hall, and the spire of the Chrysler Building.

- **Primitive arts** – The simplified sculptural forms of African, Egyptian and Aztec Mexican art and architecture influenced contemporary designers to omit inessential detail. The discovery of Tutankhamun's tomb in 1922 and subsequent exhibition sparked the world's interest in all that was ancient Egyptian and Art Deco responded with some quite literal interpretations.

- **Machine age** – The Art Deco style celebrates the machine age through explicit use of man-made materials (aluminium, glass and stainless steel), symmetry and repetition. Architecture celebrated man's technological achievements in building skyscrapers and ocean liners.

Eileen Gray (1879–1976)

Gray was born in Ireland to a wealthy family of artists and began her university career at the Slade School of Fine Arts in London as a painter. She eventually left painting to study lacquer work under the guidance of Japanese lacquer craftsman, Sugawara. In 1913, she held her first exhibition, showing some decorative panels at the Salon des Artistes Décorateurs in Paris. Here she successfully combined lacquer and rare woods, geometric abstraction and Japanese-inspired motifs into her work.

During the First World War she remained almost permanently in London and only returned to Paris in 1918. Until 1919 she worked as an independent furniture designer, and thereafter as an interior decorator. Her interior designs generated a great deal of praise in the press – amongst her admirers was Walter Gropius, the founder of the Bauhaus. In 1922 she opened the Jean Désert gallery as a showcase for her own designs.

Shortly thereafter, persuaded by Le Corbusier and her lover Badovici, she turned her interests to architecture. In 1924 Gray and Badovici began work on their vacation house, **E-1027** in southern France. This is considered to be her first major work, successfully blurring the border between architecture and decoration with a highly personalised design to fit in with the lifestyle of its intended occupants.

Figure 3.39 This 'Hollywood style' example of Art Deco architecture clearly shows the geometric forms and ocean liner aesthetics of the style.

It embraced both hand-crafted and machine production, exclusive works of high art and mass-produced products in affordable materials.

Art Deco reflected the ever widening needs of the contemporary world. Unlike the stark functionalist principals of Modernism, it responded to the human need for pleasure and escape. Art Deco was an opulent style, and its lavishness is attributed to reaction to the forced austerity imposed by the First World War. Geometric forms and patterns, bright colours, sharp edges, and the use of expensive materials, such as enamel, ivory, bronze and polished stone are well-known characteristics of this style, but the use of other materials such as chrome, coloured glass and Bakelite also enabled Art Deco designs to be made at low cost. This eclectic and elegant style soon

Figure 3.40 *Gray's E-1027 villa in France successfully blended her skills as an architect and interior and furniture designer.*

E-1027 is a codename that stands for the names of the couple: E for Eileen, 10 for Jean (the tenth letter of the alphabet), 2 for Badovici and 7 for Gray. Gray designed the furniture as well as collaborated with Badovici on its structure. Her circular glass E-1027 table and rotund Bibendum armchair were inspired by the recent tubular steel experiments of Marcel Breuer at the Bauhaus. Both pieces of furniture have become design classics and are still produced to this day.

Le Corbusier visited E-1027 on numerous occasions and admired her work very much. Unfortunately for Gray, Le Corbusier loved it so much that he was moved to add his own touch to the clean white villa, painting a series of colouful wall murals, an act which Gray considered to be vandalism.

WEBLINKS:

www.decopix.com
www.deco-world.co.uk

Figure 3.41 *Gray's Bibendum armchair designed for E-1027 is available to buy even today.*

Streamlining (1935–1955)

Philosophy

Towards the end of the Art Deco period a new style emerged known as Streamline Moderne, influenced by the modern aerodynamic designs derived from advancing technologies in aviation and high-speed transportation. This was a period of new materials and mass-production processes that could produce more refined products. It was an age when people were looking excitedly to the future and even into outer space.

Streamlining is the shaping of an object, such as an aircraft body or wing, to reduce the amount of drag or resistance to motion through a stream of air. A curved shape allows air to flow smoothly around it. Therefore, in order to produce less resistance, the front of the object should be well rounded and the body should gradually curve back from the midsection to a tapered rear section – a classic teardrop design.

Aerodynamics had been considered by designers for use with automobiles since the turn of the century but it wasn't until the 1930s that new materials and processes were available for cost-effective production. Soon both American and European industrial designers were producing experimental 'teardrop'-based concepts. Although none of these reached production, they had the effect of broadening the minds of the consumer and pointing future design in a new direction.

Figure 3.43 Perhaps one of the best-known examples of streamlining was the Volkswagen Beetle designed by Ferdinand Porsche. Its origins in streamlining are apparent and it went on to become the most popular and longest-produced automobile ever.

These attractive teardrop shapes were enthusiastically adopted within Art Deco, even applying streamlining techniques to domestic appliances such as radios, vacuum cleaners and refrigerators. Although efficient aerodynamics are not a key feature of many products, the combination of streamline form and modern materials made them stand out from their competitors and therefore more appealing to a growing consumer society.

In the 1950s came the 'Space Age' and the 'Atomic Age' and with it the 'Googie' style of architecture. Googie epitomises the spirit of a generation looking excitedly towards a bright, technological and futuristic age, characterised by space-age designs that depict motion such as boomerangs, flying saucers, atoms and parabolas.

Style

- **Teardrop shape** – With the sleek, efficient forms of airliners and marine life as inspiration, the form adopted as perfect aerodynamicism was that of the teardrop; with the round end being the front. This aerodynamic form became the new aesthetic direction and guided the design of modern products.

- **Futuristic design** – Science fiction provided optimism for a new and better future with sleek rocket shapes and atom designs.

Raymond Loewy (1893–1986)

Loewy was one of the best-known industrial designers of the 20th century. Born in France, he spent most of his professional career in the United States where he influenced countless aspects of American life from transportation to commercial art.

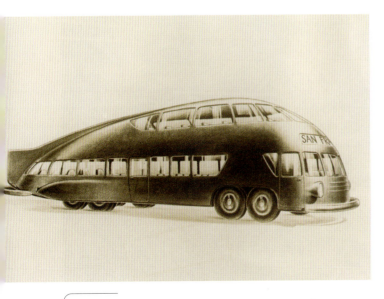

Figure 3.42 Norman Bel Geddes patented designs for a teardrop car, bus, yacht, liner and plane although they were never commercially produced.

Loewy launched his career in industrial design in 1929 when a British manufacturer of duplicating machines commissioned him to improve the appearance of one of their products. In three days, the 28-year-old Loewy designed the shell that was to encase these duplicators for the next 40 years. In doing so he was the first designer to transform the look of a product by streamlining, which he referred to as "beauty through function and simplification".

Loewy is often referred to the 'father of industrial design' as it was he who first promoted the idea that large corporations could hire industrial designers to provide outside advice on the development of their products. He demonstrated the practical benefits derived from his functional styling, stating:

"Success finally came when we were able to convince some creative men that good appearance was a saleable commodity, that it often cut costs, enhanced a product's prestige, raised corporate profits, benefited the customer and increased employment."

During the 1930s Loewy established a relationship with the Pennsylvania Railroad, and his most notable designs for the firm were their streamlined passenger locomotives. The GG-1 electric locomotive demonstrated on an even larger scale the efficiency of industrial design. The welded shell of the GG-1 eliminated tens of thousands of rivets, resulting in improved appearance, simplified maintenance, and reduced manufacturing costs. As the first welded locomotive ever built, the GG-1 led to the universal adoption of the welding technique in their construction.

During this period Loewy also began a long and productive relationship with US car-maker Studebaker, producing the iconic bullet-nosed cars as well as modernising their logo design. Loewy's car designs incorporated new technological features; introducing slanted windshields, built-in headlights and wheel covers in his designs, he also advocated lower, leaner and more fuel-efficient cars long before fuel economy became a concern. He was still designing for Studebaker in the 1960s when they launched his most successful car: the 'Avanti' (Italian for forward).

Figure 3.44 *Loewy's aerodynamic streamlining for high-speed locomotives.*

As a commercial artist he is credited with designing the classic 'Lucky Strike' cigarette packaging. By changing the package background from green to white, he reduced printing costs by eliminating the need for green dye. In addition, by applying the red Lucky Strike target on both sides of the package, he successfully increased the product visibility and, ultimately, product sales. Other successful projects included the design of the Shell, Exxon and BP logos for the petroleum giants.

From 1967 to 1973 Loewy was commissioned by NASA as a habitability consultant for the Saturn–Apollo and Skylab projects where he helped design the interior living spaces for spaceships. Proven time and again, Loewy's design principles continue to be relevant years later.

WEBLINKS: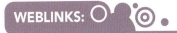

www.raymondloewy.com
www.raymondloewy.org

Post-modernism (1975–Present)

Philosophy

The term 'post-modernism' was first coined by architect Charles Jencks. He used it to criticise the functionalism of the Modernism movement and to describe the eclectic new design styles being developed by a whole range of contemporary architects and designers. The debate regarding whether the term post-modernism, meaning 'after modernism', is appropriate still rages to this day as it does not seem to encompass the range of contemporary thinking and design styles. Indeed, to many the modern movement has not ended as a lot of its ideals are still in use today.

The movement of Post-modernism began with architecture, as a reaction against the perceived blandness and hostility present in modernist architecture as preached by the Bauhaus. Its philosophy of an ideal perfection, harmony of form and function and dismissal of decoration was at odds with contemporary designers who wanted individualism and personality back into design.

Figure 3.45 *Richard Rogers' Pompidou Centre in Paris received mixed reviews when it opened in 1978. Its visible support structure and service pipes are a complete contrast to modernist minimalism.*

Out of this period came the the Memphis group comprising Italian designers and architects who created a series of highly influential products in the 1980s. Founder member Ettore Sottsass disagreed with the approach of the time and challenged the idea that products had to follow conventional shapes, colours, textures and patterns. They drew inspiration from such movements as Art Deco and Pop Art, styles such as the 1950s Kitsch and futuristic themes. Their concepts were in stark contrast to so-called 'good design'. On the launch of the Memphis furniture group in 1981 Sottsass challenged conventional taste by stating:

"Every journalist reacted by saying that the furniture was bad taste. I think it's super taste. It is Buckingham Palace that is bad taste. Memphis relates to the actual world; we are quoting the present, and the future."

The work of the Memphis Group has been described as vibrant, eccentric and ornamental. It was conceived by the group to be a 'fad' that, like all fashions, would very quickly come to an end. In 1988, Sottsass dismantled the group.

New Design style

- **Humour and personality** – Products were bright and colourful like children's toys, often including unnecessary decoration in an attempt to give static objects personality. By providing products with personality it made them more appealing to the consumer who wanted to express their individuality.

- **'Retro' design** – Designs that take inspiration from past movements and styles and re-interpret them in a modern way. Alternatively, the copying of old designs but manufactured from modern materials and incorporating modern technology to satisfy the trend in nostalgia.

- **Deconstruction** – A development in architecture where the surface structure of a building is distorted so that it becomes non-rectangular. The finished visual abstract and non-symmetric appearance gives the impression of controlled chaos.

Figure 3.46 *Sottsass' Carlton Bookcase combines his interest in Indian and Aztec art with 1950s popular culture to produce a bright, colourful and shocking style.*

Philippe Starck (1949–)

Philippe Starck is a well-known French designer and probably the best-known designer in the New Design style. His designs have been well publicised in the media and include a diverse range from spectacular interior designs to mass-produced consumer goods such as toothbrushes, chairs and even houses.

Starck has worked independently as an interior designer and as a product designer since 1975. He rose to fame in 1982 with his interior designs for the French President's private apartments. Since then he has collaborated with many multi-national companies on the design of packaging and relatively inexpensive products such a mouse for Microsoft.

Starck's products are often stylised, streamlined and organic in their appearance. They posses humour and he often christens his products with names to bring them to life and give each an individual personality. He values new technologies and has always possessed a taste for innovation with a conviction that "it's better to make a creative mistake than a stagnant work in good taste". He is also concious of sustainability and products are often light and economical in their use of energy and materials, from production to consumption, packaging and transportation.

Starck's work for the Italian company Alessi has produced some of the best-known icons of the late 20th century.

LINKS TO:

Unit 3: Design for the Future: Form and function – another classic Alessi product designed by Starck is the kettle 'Hot Bertaa', which is discussed in the next section.

WEBLINKS:

www.starck.com
www.alessi.com

Figure 3.47 Starck considers himself as 'a Japanese architect, an American art director, a German industrial designer, a French artistic director, an Italian furniture designer'.

Figure 3.48 This sleek lemon squeezer named the 'Juicy Salif' was created for Alessi in 1990 and has since become an affordable and popular cult item.

FACTFILE:

Design movements

Movement	Philosophy	Style	Designers	Visual reference
Arts and Crafts	• Fitness for purpose. • Honesty in design and making, the return to the designer– craftsman as a reaction against industrialisation.	• Simplicity. • Natural forms and materials.	William Morris Ernest Gimson C.R. Ashbee	
Art Nouveau	• The languid line. • The formulation of new aesthetic values for a new urban lifestyle.	• Curvy 'whiplash' lines and stylised flowers. • Languid feminine form.	Charles Rennie Makcintosh L.C. Tiffany Antoni Gaudi	
Bauhaus Modernism	• Functionalism. • Reducing form to the most essential elements by omitting decorative frills.	• The machine aesthetic using modern materials. • Simple, geometrically pure forms and clean lines.	Walter Gropius Marcel Breuer Ludwig Mies van der Rohe	
Art Deco	• Popular modernism. • Opulent architectural and decorative arts style in reaction to post-war austerity.	• Zig-zagged, geometric fan motifs. • Symmetry and repetition. • Inspiration from ancient Egypt.	Eileen Gray Albert Anis Walter Dorwin Teague	
Streamlining	• Consumerism and style. • New prosperity and widened consumer choice. • Celebrating speed and efficiency.	• Aerodynamics. • Teardrop shape. • Futuristic inspiration.	Raymond Loewry Norman Bel Geddes Henry Dreyfuss	
New Design style (Post-modernism)	• 'Less is a bore!' expressive and individual as opposed to modernist functionalism.	• Humour and personality • Retro design • Deconstruction	Phillipe Starck Richard Rogers Ettore Sottsass	

3 Form and function

The connection between form and function has been one of the most controversial issues in the history of design. When products were first mass produced in Victorian times they were highly decorated to look like hand-made products, whether their decoration was appropriate or not. The development of 'reform' groups such as the Arts and Crafts movement gradually brought about change in the concept of design. The form of products was to be simplified and the products well made from suitable materials. At the turn of the 20th century, developments in materials and technology enabled the production of innovative new products such as the telephone. Many of these products were so innovative that there was no benchmark on which to base their designs.

The development of mass production techniques required that products be standardised, simple and easy to produce. The modernist movement, which supported functionalism, suggested that the form of a product must suit its function and not include any excessive or unnecessary decoration. Therefore, for a product to be mass produced at a profit, it needed to be simple and easy to produce.

FACTFILE:

'Form versus function'

'Form follows function'	Functionality as the prime driver	Functionalists support the view that products should be fit for purpose without any unnecessary decoration.
'Form over function'	Aesthetics as the prime driver	Supporters advocate the aesthetic qualities of products in contributing towards an overall aesthetically pleasing environment.

For many consumers these days, design has become an important means of self-expression. Consumers choose products not just for what they do, but for what they tell the world about their lifestyle choices. Many products are no longer simple, functional artefacts. For example, the purchase of a pair of trainers takes into account many factors such as how comfortable they are but the overriding reason for buying them may be their appearance and branding. Product performance and reliability are no longer real issues for the consumer as most products carry guarantees and are subject to rigorous QA procedures. The main reason for choosing one product from another with similar functions is its aesthetic qualities.

One of the roles of the designer, then, is to provide the product with the right style or image for a particular target market group. Get this wrong and the product will not sell; get it right then it will become an 'object of desire' for aspiring consumers. Now that so many products are mass produced and sold in their millions, the designer must inject a sense of individuality or personality into an object. For example, the Italian design company Alessi is famous for its playful design of affordable objects and appliances for the kitchen, using bright and colourful polymers and stainless steel to create contemporary and humorous products. However, Alessi did overstep the line between form and function with the Hot Bertaa kettle designed by Phillipe Starck, which it had to withdraw from production as it didn't boil water very well or safely. This is an important lesson for any modern designer, who must strike a balance between the form and function of the product.

Table 3.18 Comparison of form and function of two kettles.

'Form over function'	'Form follows function'
Philippe Starck's Hot Bertaa kettle	High street 'jug' kettle

With reference to form	
• Form over function where kettle is a sculptural 'object of desire' and a lifestyle product. • Aesthetically pleasing for fashion conscious incorporating a brightly coloured polymer handle, aluminium body and cone-shaped shaft that pierces the body of the kettle serving as both its handle and spout.	• Form follows function where the function of boiling water is the prime driver and the secondary requirement is to look good in the kitchen. • Inoffensive, neutral style that fits in with a wide range of domestic kitchen environments. • Attractive to wide range of customers with curved handle with ergonomic grip, stainless steel body with contrasting blue water level indicator.

With reference to function	
• Art form where functionality became irrelevant – does not have to look like a kettle in order to boil water. • Poor functionality and user friendliness as the shape proved difficult to pour boiling water from and impractical to fill water through the narrow cylindrical handle. • Intrinsic design flaws and poor safety as heating mechanism largely unreliable, handle became hot once water had boiled, water leaked easily and dangerous to lift as it weighed so much.	• Good functional aspects and user friendliness including an ergonomically designed handle grip for comfortable pouring, an ON/OFF switch positioned at top for easy access, the handle at the side of the kettle body for easy filling and pouring, a water level indicator tells the user how much water is in kettle and a large opening lid to fill the kettle. • Important safety features incorporated such as automatic switch off once boiled and the kettle is removable from power supply base so no risk of trailing power cable.

THINK ABOUT THIS!

What, in your opinion, is more important – form or function? Can you have one without the other? Or would the ideal product strike a balance between the two? Select a range of similar products, such as trainers or MP3 players, and compare them with reference to both form and function.

4 Anthropometrics and ergonomics

The relationship between anthropometrics and ergonomics

Ergonomics

Ergonomics is the science of designing products, systems and environments for human use. This means applying the characteristics of human users to the design of a product – in other words, matching the product to the user. Ergonomics is therefore an essential part of the design process. Sometimes products are matched to a single user, where the product is customised to suit one person. The main objective of ergonomics to the designer is to improve people's lives by increasing their comfort and satisfaction when interacting with products, systems and environments. In order to achieve this, data about the size and shape of the human body is required – this branch of ergonomics is called anthropometrics.

Anthropometrics

Anthropometrics is a branch of ergonomics that deals with human measurements, in particular their shapes and sizes. For many products, systems and environments, complex data is required about any number of critical dimensions relating to the user, such as height, width or length of reach when standing or sitting. Therefore, anthropometric data must take into account the greatest possible number of users. This data exists in the form of charts that provide measurements for the 90 per cent of the population that falls between the fifth and 95th percentiles (see Figure 3.49).

Sources and applications of anthropometric data

When applying anthropometric data to a design problem, the designer's aim is to provide an acceptable match for the greatest possible number of users. This is achieved by the use of data charts such as those issued by the British Standards Institute (BSI), which are available in a simplified form from the *Compendium of essential design and technology standards for schools and colleges*. Simple data charts relating to measurements for men, women and children can also be found in the clothing sections of mail order catalogues.

Statistical data available from the BSI is associated with heights of men, women and children. The height at which 5 per cent of the population is shorter is known as the fifth percentile. Likewise, the height at which only 5 per cent of people are taller is known as the 95th percentile. The anthropometric data that covers 90 per cent of the British population covers those who fall between the fifth and 95th percentiles. Putting this to use, a furniture designer would have to take into account a range of heights from 1.5 metres to 1.9 metres when designing a chair that would be comfortable for all users. According to the principles of anthropometrics, the designer would ignore the smallest (5 per cent) and tallest (5 per cent) users and design the chair to fit the remaining 90 per cent, who account for the greatest number of users.

Anthropometric data will vary for different regions in the world. For example, the average height of people in Japan is shorter than in the UK. Conversely, the average height of people in Scandinavian countries is taller than in the UK. When designing for the disabled, specialist data is available that takes into account wheelchair use, for example. There are also niches in the market for products and clothing designed specifically for short and tall people with specialist companies and stores.

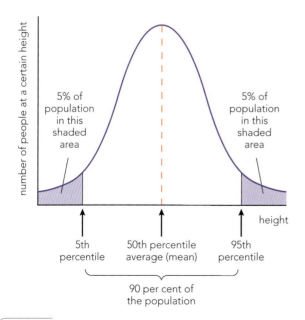

Figure 3.49 *Anthropometric data incorporates human measurements representative of 90 per cent of the population (the fifth to 95th percentiles).*

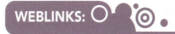

WEBLINKS:

www.bsi.org.uk/education

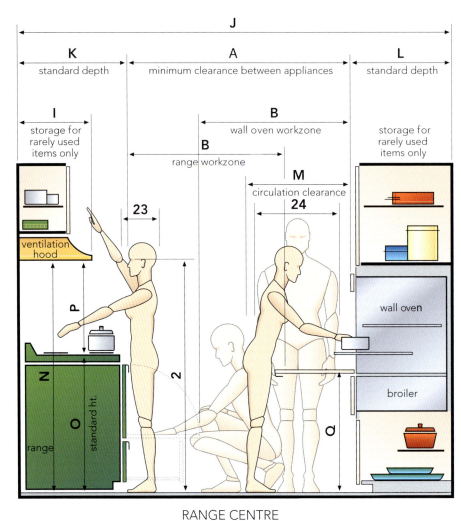

RANGE CENTRE

Figure 3.50 An example of anthropometric data used for kitchen planning.

Key ergonomic factors for the designer

The study of key ergonomic factors by designers is used to improve the design of products, systems and environments by understanding and/or predicting how humans interact with their surroundings. The methods used focus upon different aspects of human performance and take quantitative and qualitative approaches.

- A **quantitative** (measurable) approach predicts the physical fit of the product to the human body shape, encompassing workload, speed of performance and errors.

- A **qualitative** (opinion-based) approach predicts user comfort and their satisfaction with the product, so it has the optimum interaction with the user.

Designers can respect the diversity of human shapes and sizes, making products suitable for both the largest and the smallest people, in four different ways.

- A single design that is valid for everybody: for example, making doorways wide enough for anyone to pass through them, regardless of their body size, the fact that they are carrying something or that they are in a wheelchair.

- Designing a range of objects that covers all possibilities, for example, in clothes sizes.

- Designing a product that is adaptable to different dimensions, for example, a chair whose height can be adjusted.

- Designing an accessory that adapts itself to an original design, for example, car seats for children.

When designing any product it is very important to study the target market group (TMG). The profile of the end-user can often help when designing a product. To this end, ergonomics draws on a wide variety of disciplines including anatomy, physiology, biomechanics, psychology, design and ICT to make beneficial changes to work and leisure activities to best fit the person carrying out the task. This improves physical and mental health, safety, productivity and efficiency. The following problem was encountered by designers at the Universal theme park in Orlando, USA:

The problem: *Shrek* is a costumed character that performs daily in the theme park. The performers that bring *Shrek* to life have had issues with shoulder, neck, and lower-back pains. Due to a number of back, neck and shoulder injuries sustained by the performers who wear the costume, a redesign of the costume was necessary to prevent further injury. The issues identified were weight of the costume, body positioning in the costume, ventilation and visibility. This resulted in 21 first-aid cases, 26 recordable injuries, 622 light duty days and 49 restricted/lost workdays from June 2004 to March 2006 with a direct cost of over US$80,000.

The solution: to achieve the weight reduction, the heavy harness was replaced with lighter materials and additional non-essential layers were removed to reduce it by 10lbs. To straighten body and head positioning, the head was separated from the body to allow movement and to balance weight in line with the spine. A ventilation opening was added into the back of the head to allow for air-flow assisted by a built-in battery-powered fan. The separation of the head, including added mobility to move the neck, resulted in increased visibility. The solution cost just US$100 in materials and US$1,600 for labour and installation – a low-cost, high-yield solution.

The outcome: after changes were implemented on March 1st, 2006 the effects were impressive: injuries fell to zero, with no lost time. The result was increased performance, zero injuries and a simple solution for a low cost of US$1,700, all due to designers listening to the needs of the user instead of making the person adapt to the job.

Adapted from The Ergonomics Society, 15th May 2007

WEBLINKS:

www.ergonomics.org.uk

Figure 3.51 *Even ogres need an ergonomic makeover!*

The interaction between users, products, systems and environments

All types of products, systems and environments are designed so their dimensions suit those of their end-users. Products need to be designed so that they can be operated easily and safely by 90 per cent of the population. Safety considerations require easy operation of products but must also ensure that the product is not operated by mistake.

When designing complex products, systems and environments such as vehicles, designers need to take account of a wide range of ergonomic considerations and anthropometric data covering the different heights and reaches of both male and female drivers.

Table 3.19 A driver's interaction with a car interior.

Interface	Ergonomic consideration
Driver's seat	• Seat adjustable backwards and forwards to take into account different leg and arm reaches in relation to steering wheel and foot pedals. • Seat adjustable up and down to take into account different heights so as to see clearly over dashboard. • Shape of seat to take into account different body sizes and padding for comfort. • Seat adjustable for comfort when driving long distances i.e. lumbar support. • Adjustable head rest to support neck on long journeys.
Steering wheel	• Diameter of steering wheel to aid control of turning. • Thickness of steering wheel, use of soft materials and textures/pistol grip features to provide comfort and safe grip. • Adjustable steering wheel (up and down) to take into account driving styles.
Gear stick/ handbrake	• Positioning in relation to driver's seat. • Diameter of gear knob and handbrake, use of soft materials and textures to provide comfort and safe grip.
Foot pedals	• Distance from seat (adjustable backwards and forwards). • Size and spacing between pedals to take into account different shoe sizes. • Texture on pedal to provide a non-slip surface.
Dashboard instrumentation	• Layout of the dashboard so that all instrumentation is within easy reach of driver. • Essential instrumentation such as speedometer is easily seen (mostly through the steering wheel). • Instant visual impact of instrumentation, i.e. warning/safety lights. • Adjustable brightness of illumination of instrumentation when driving at night. • Size and shape of knobs and switches for ease of use, limiting the need for hand to be removed from steering wheel.
Heating/air conditioning	• Temperature control of the car's interior environment is important for comfort.
Mirrors	• Mirrors that can be fully adjusted from the driver's seat to provide all-round high visibility, i.e. ball joint for manual adjustment of rear view mirror, electric wing mirrors.

LINKS TO:

Ergonomic considerations when designing computer workstations in **Unit 2: Health and Safety:** carrying out risk assessments in accordance with the HSE for the design of graphic products using computers and manufacture of models and prototypes using workshop practices.

Sustainability

Getting Started!

Sustainability means safeguarding the world for ourselves and for future generations, using energy and other resources in a way that minimises their depletion, and designing for a better quality of life. In recent years we have had to rethink our approach to design, materials usage and manufacturing methods by moving towards an approach that considers the economic, social and environmental issues and the use of cleaner design and technology. These are not simply issues for governments and large companies – how can *you* contribute towards a more sustainable future?

1 Life-cycle assessment

It is the case with any design decision and solution that an optimum is looked for and a balance drawn between cost and benefit. Balancing the needs against the impact to the environment is becoming increasingly more difficult for manufacturers as they strive to develop new products and processes. Life-cycle assessment (LCA) is a technique now widely used to assess and evaluate the impact of the product or packaging 'from the cradle to the grave' through the extraction and processing of raw materials, the production phase, and life-cycle processes including distribution, use and final disposal of the product.

Life-cycle inventory

Consumers are becoming increasingly aware of environmental issues and expect companies to pay attention to the environmental impact of their products. In addition, the BSI and the ISO 14000 series of standards now demand continuous improvement in a company's environmental performance. In order to help measure this, many companies now use a life-cycle inventory. A life-cycle inventory describes which raw materials are used and what emissions will occur during the life of a product. The basis of this study is to compile a list of all the inputs and outputs of industrial processes that occur during the life cycle of a product, including:

- **environmental inputs and outputs** of raw materials and energy resources
- **economic inputs and outputs** of products, components or energy that are outputs from other processes.

The life-cycle inventory can be expressed as a process tree (see Figure 3.53 below) where each box represents a process with defined inputs and outputs that forms part of the life cycle. The second stage of this process is to interpret the data to assess the overall environmental impact of the product.

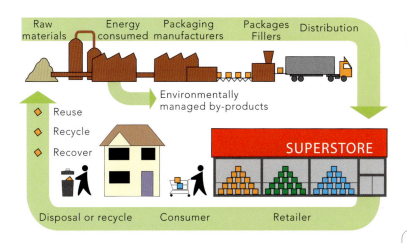

Figure 3.52 Life-cycle assessment (LCA) for packaged products.

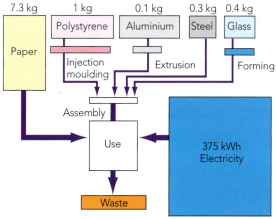

Figure 3.53 This simple process tree of an LCA inventory for a coffee machine clearly indicates design priorities: minimise the use of electricity and paper filters.

2 Cleaner design and technology

Sustainable product design

Successful cleaner design requires reducing the environmental impact of the product or packaging throughout its entire life cycle. Issues such as raw material use, waste production, energy consumption and emissions into the atmosphere all need to be considered at each stage of the product life cycle. It is now fundamentally important for a designer to also consider *design for recycling*, including the following requirements of a product:

- easy to dismantle for repair or re-use and so extending product life

- easy to separate different materials for recycling

- easy to remove components that must be treated separately for repair

- use as few different materials as possible

- mark the materials/polymers in order to sort them correctly

- avoid surface treatments in order to keep the materials 'clean'.

In his book *The Total Beauty of Sustainable Products*, Edwin Datschefski spells out five factors that make up product sustainability.

1. **Cyclic:** products that are made from biodegradable organic materials or from minerals that are continuously recycled, decreasing levels of waste and pollution; for example, products made from Biopol®.

2. **Solar:** products that in manufacture and use consume only renewable energy that is cyclic and safe; for example, products made using renewable energy sources such as wind power and products that operate using solar-powered (photovoltaic) cells.

3. **Safe:** all releases to air, water, land or space are 'food' for other systems; for example, products that do not emit unnecessary pollutants or chemicals during their manufacture.

4. **Efficient:** products that in manufacture and use require 90% less energy, materials and water than equivalent products did in 1990; for example, a reduction of materials used in packaging a product, which decreases the amount of raw materials extracted, energy used in processing/manufacturing, and pollution, etc.

5. **Social:** products whose manufacture and use supports basic human rights and natural justice; for example, Fairtrade products that help producers in developing countries receive a fair share of profits, which reduces exploitation of the workforce.

FACTFILE:

Key environmental considerations for cleaner design and technology

Life-cycle stage	Key environmental considerations
Raw materials	• Use less material. • Use materials with less environmental impact. • Consider recyclable materials. • Adhere to relevant legislation.
Manufacture	• Reduce energy use. • Simplify processes where appropriate. • Reduce waste. • Use natural resources efficiently.
Distribution	• Reduce or lighten packaging. • Reduce mileage of transportation to customer.
Use	• Increase durability of product. • Encourage refill consumables where appropriate. • Use 'green' credentials as a positive marketing strategy. • Promote efficient use of product.
End-of-life	• Make reuse and recycling easier. • Reduce waste to landfill.

Raw materials

The key issues for designers when considering the use of materials for products and packaging are the environmental and economical costs of the raw material. Although metals are abundant in the Earth's crust, their extraction and processing is costly both environmentally and financially, largely due to the vast amounts of energy required to convert the ore into the finished drinks can, for example. Polymers are derived from crude oil, which is a finite resource and for this reason designers must consider

Table 3.20 *Environmental impact of raw materials for packaging.*

Packaging material	Raw material	Extraction	Processing
Paper and board	Trees	Deforestation, environmental degradation of forest areas, distribution, etc.	Chemical pollutants used in chemical woodpulp production and bleaching.
Metals	Aluminium – bauxite ore Steel – iron ore	Environmental impact of mining activities, e.g. energy use, open-cast mining, transportation, etc.	Vast amounts of energy required to process ores and resultant carbon dioxide emissions.
Polymers	Crude oil	Environmental impact of drilling activities, e.g. energy use, destruction of habitat, etc.	Vast amounts of energy required to refine oil and produce polymers, with resultant carbon dioxide emissions.

the use of recycled materials to reduce consumption. The UK relies heavily upon imported timber and chemical woodpulp, which has to be transported long distances, resulting in high transport costs and carbon dioxide emissions.

The answer, then, is relatively simple: reduce the amount of materials used in order to conserve resources, which will in turn reduce energy consumption and pollution, and use more recycled materials or use materials that are recyclable. Of course, with current mass production and mass consumerism the solution is not an easy one.

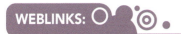

WEBLINKS:

www.incpen.org – The Industry Council for Packaging and the Environment (INCPEN)

Manufacture

The conversion of raw materials into finished products or packaging incurrs considerable environmental impact and costs. For many companies the analysis of existing manufacturing processes can identify areas that can be modified to achieve more efficient and cleaner processes. The aim is to reduce production costs by creating designs that use less material and less energy during manufacture and to reduce waste production. Modifying the design and manufacture to increase efficiency may involve using:

- a simpler design with fewer components to reduce materials use and assembly time
- different materials to reduce their weight or the quantity used

- materials that use less energy during manufacture and produce less waste
- simpler components that are easier to machine or mould and produce less waste
- a simplified or different work flow with improved quality control.

The Coca-Cola '202' drinks can

Coca-Cola Enterprises Ltd. produces and distributes around two billion soft drinks cans per year. In the early 1990s, many people switched to plastic bottles, putting pressure on can manufacturers to reduce costs. As further lightweighting was not thought possible with the existing '206' can design, a new '202' can was developed. It had a reduced end diameter, whilst maintaining the same body diameter to contain the same volume of liquid. The change in design had to take account of the can manufacture, the filling process, the can strength and stackability for distribution. Coca-Cola secured the agreement of European can manufacturers to a common specification for the '202' can end and body. Previous to this agreement machinery had to be reset to accommodate different suppliers' cans and different end profiles.

The new '202' can design successfully reduced raw materials costs and simplified the production processes, resulting in:

- cost savings of over £1 per thousand cans and reduced metal use world-wide in the canned drinks industry
- savings of around £2.3 million a year from 1995 onwards for Coca-Cola
- the use of lightweight cans enabling more products to fit in a single lorry, reducing the number of journeys in distribution.

Figure 3.54 *The Coca-Cola '202' can is a good example of product lightweighting.*

The possibility of reducing the amount of materials used to manufacture a product is often found in processes that involve cutting and stamping shapes from sheet materials. For example, in can manufacture careful calculations and efficient lay planning must be made to limit the amount of aluminium used for making the circular tops of the cans. There are two methods of arranging the can top on a rectangular sheet: with the circular tops in either a square or triangular formation. When the wastage of materials from the two methods is calculated it works out that the

square method produces 21.4 per cent scrap with the triangular method producing just 9.3 per cent scrap. This scrap material can be recovered and recycled. Clearly efficient lay planning has an enormous impact upon reducing material wastage.

Distribution

There are a number of issues relating to cleaner distribution of goods around the UK but they all result in the same key concerns: extremely large energy use and resultant carbon dioxide emissions, which contribute towards global warming. Congestion on our roads and motorways is increasing and road haulage companies are significantly adding to this. Other forms of transport could be used that are less polluting, such as trains (especially electric trains) or even waterways where appropriate. The size of journeys could be reduced from the manufacturer to the consumer either by use of local resources or geographical locations of distribution centres. If a lorry has to make a journey then there are a number of things that could be done to save fuel, such as reducing or lightening the amount of packaging used in products, driving sensibly and smoothly and exploring alternatives to fossil fuels.

Alternatives to fossil fuels

Apart from the obvious pollution from traditional fossil fuels, the financial cost of diesel and petrol will continue to rise. Therefore, the only realistic course of action for drivers and transport companies is to find less polluting and cheaper alternatives that address three key factors: good performance, reliability and availability. Unfortunately, it is the lack of availability of these alternative fuels that is the main reason they are not widely used at present.

WEBLINKS:

www.dft.gov.uk/ActOnCO2 – Department for Transport website devoted to cutting carbon dioxide emissions

www.toyota.com/prius – information on the Toyota Prius hybrid electric vehicle

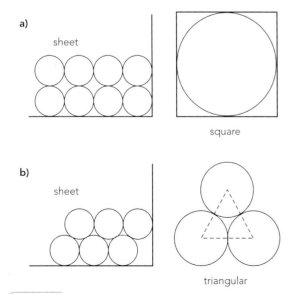

Figure 3.55 *Efficient lay planning reduces material wastage.*

Table 3.21 *Advantages and disadvantages of alternative fuels.*

Fuel type	Advantages	Disadvantages
Liquefied petroleum gas (LPG)	• Relatively good fuel availability. • Good range of kits available. • Reduced emissions. • Increasingly good supply of used vehicles. • Low-cost fuel – less than 50% of diesel. • Reliable performance.	• Not available for diesel vehicles. • No factory-fit models available.
Bio ethanol	• Reduced emissions. • Increased power. • Factory-fit models now available. • Renewable fuel.	• Very poor availability of fuel. • Limited availability of vehicles. • Similar price to diesel. • Up to 30% lower economy than petrol.
Compressed natural gas	• Kits fit to existing diesel vehicles, such as HGVs. • Similar economy to diesel. • Reduces diesel emissions.	• Very poor availability of fuel. • Limited availability of kits and vehicles. • Slow refuelling times.
Hydrogen	• Zero emissions. • Renewable fuel.	• Very poor availability of fuel. • Limited availability of kits and vehicles.
Electricity	• Zero emissions.	• Very limited range. • Slow charging/refilling time.

Figure 3.56 *The Toyota Prius was one of the first mass-produced hydrid electric vehicles that switches between an electric motor for city driving and an efficient petrol engine for the open road.*

Use and maintenance

The designer's responsibilities do not end after the product reaches the consumer – reducing the impact from the use of the product must also be considered. Many products are designed and manufactured in such a way that it makes it virtually impossible to access internal components if something stops functioning. This 'built-in obsolescence' means that the product cannot be repaired and therefore has to be discarded and replaced. The issue here, then, is one of repair versus replacement. For example, many Seinnheiser DJ headphones have replaceable parts for extended product life. The parts that are most likely to become damaged, such as the ear pads (tearing) or the cable (being pulled out), can be ordered individually from the company and are easily replaced.

There is also the issue of upgrading technology when it becomes obsolete. For example, traditional television sets that receive analogue signals will soon become obsolete when digital transmission supercedes the old analogue form. Most people will rush out and buy a new flat screen digital model even though there is plenty of life left in the old one. But, analogue televisions can be converted to receive digital signals directly; they may not look as good as the new trendy flat screen versions but they are much more environmentally friendly.

Figure 3.57 *These Seinnheiser HD 200-V1 DJ headphones have replaceable parts for extended product life.*

THINK ABOUT THIS!

Have you ever purchased a product and asked whether parts could be replaced if they fail? Why do you suppose that this is not a major issue for many people and how could the repair and not replacement of products be encouraged? Have manufacturers got to encourage this or have the attitudes of society got to change?

LINKS TO:

Unit 3: Design for the Future: Design in context: The effects of technological changes on society: built-in obsolescence.

3 Minimising waste production

Perhaps the most important economic factor for a designer of sustainable products to consider is that waste is lost profit. There are some simple options to consider when deciding how to minimise waste production at the end-of-life stage; they are referred to as the four Rs:

- reduce
- reuse
- recover
- recycle.

Reduce

For all designers, one of the first priorities for sustainability should be to reduce the quantities of any material chosen whenever possible. Therefore, packaging designers must optimise the amount of materials needed to package a product in order to minimise the consumption of resources, which will in turn achieve significant cost savings and improve profit margins.

Manufacturers are obliged to reduce packaging use under the UK Producer Responsibility Obligations (Packaging Waste) Regulations 1997. The Government's Envirowise programme suggests that manufacturers:

- consider the materials and designs they use

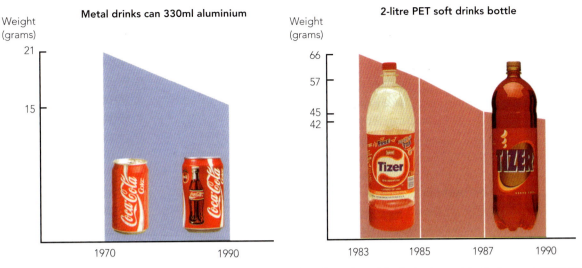

Figure 3.58 *All types of packaging materials have evolved to contain the same volume of goods with less weight of material.*

Source: The Industry Council for Packaging and the Environment (INCPEN)

- examine ways of eliminating or reducing the packaging requirement of a product – changes in product design, improved cleanliness, better handling, JIT delivery, bulk delivery, etc.
- optimise packaging use, i.e. match packaging to the level of protection needed.

In one Envirowise case study, Ambler of Ballyclare implemented an environmental policy focusing upon the minimisation of packaging waste. The financial benefits were considerable (£103,002 savings per year), as were the environmental benefits including diverting waste from landfill, reducing environmental impact and reducing disposal costs. The reduced use of vehicles for the transportation of waste also saw a 42% reduction in carbon monoxide emissions.

WEBLINKS:

www.envirowise.gov.uk – independent advice and support on practical ways to increase profits, minimise waste and reduce environmental impact for UK businesses.

Figure 3.59 *Systematic approach to identifying packaging use and waste minimisation by a company.*

Examine and monitor each type of packaging used at Ambler

Is the use necessary? — **Yes** / **No**

No → Elimination
Achieve elimination of:
- polythene bags on light-coloured yarn
- polythene wrapping on waste bales*

Can packaging be reduced without affecting product quality? — **Yes** / **No**

Yes → Reduction
- use of one-trip cardboard boxes

Can packaging be reused or modified for reuse? — **Yes** / **No**

Yes → Reuse
On site
- reusable cartons
Off site
- woven polypropylene packaging

Can packaging be recycled? — **Yes** / **No**

Yes → Recycling
Consider recycling outlets for:
- PET strapping
- polythene packaging on incoming fibre
- waste paper and cardboard

No → Minimisation
Consider minimising waste by:
- minimising waste to landfill
- compacting waste to reduce transport costs**
- practising good housekeeping to minimise general waste

* Involved purchase of baler
** Involved purchase of compactor

147

Reuse

Part of the cyclic factor of sustainable design is the reuse of products, which minimises the extraction and processing of raw materials and the energy and resources required for recycling. A number of companies adopt returnable or refillable containers for some of their products; for example, the door-step delivery of milk in glass bottles. Refillables appear to offer environmental benefits yet they often require greater use of resources in their manufacture and distribution to enable them to withstand the rigours of repeated use. This initial use of extra resources can be offset by the reuse of the container but only in local distribution and collection schemes. If reuse is to be economically viable then the cost of collection, washing and refilling should be less than producing a new container.

Refillable containers have been one of the most dramatic developments in retailing in recent years, most notably in the reduction of packaging size of detergents and fabric conditioners. Concentrated forms of these products have been realised through technological developments, which has resulted in less packaging per dose of detergent.

Recover

The manufacture of any product obviously requires the use of energy. If the product is simply discarded and landfilled then all of this energy is lost. Waste that cannot readily be recycled but can burn cleanly can be incinerated in specialised power stations to generate electricity and provide hot water for the local area. This is not an ideal solution (waste reduction is), but by adopting such technology less finite fossil fuel is needed to generate electricity in conventional power stations. In Sweden 47% of waste is recovered in energy from waste plants and in the Netherlands 34% is recovered. Tetra Pak education service has claimed that you could run a 40W bulb for an hour and a half on the energy released when one aseptic carton is burnt.

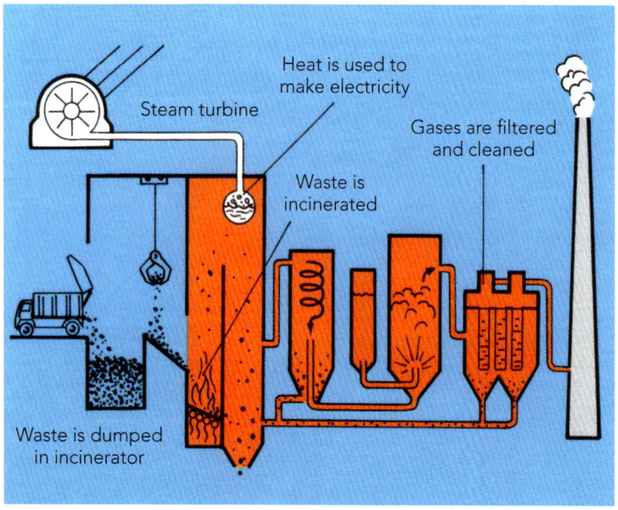

Recycle

Essentially, recycling takes waste materials and products and reprocesses them to manufacture something new. Some materials, such as paper and boards, can be made into the same products and others can be made into something completely different such as plastic vending cups into pencils. Recycling is an important aspect of a modern consumer society with millions of tonnes of waste being disposed of in landfill sites or incinerated, causing environmental concerns.

Recovery and recycling of paper and board

The cellulose fibres found in a variety of plant life can be used to manufacture paper and board – and it can generally be reused between four and six times. New fibres can be processed in pulp mills to produce wood pulp for virgin paper manufacture. This virgin paper and board is of a high quality and strength due to its content of new fibre, so is used for specialist applications. However, almost any type of paper and board can be recycled using an elaborate grading and sorting system that ensures maximum value is obtained from the recovered fibre. As these recycled fibres are of a lower quality – weaker and maybe contaminated – the products they are made into are designed to take into account the particular characteristics of recycled fibre.

Corrugated board – can be made from recycled paper for single or multi-ply kraft liners (flat sheets) and fluting (corrugation) provided liquid starch is added to provide stiffness and extra strength. UK manufacture of materials for corrugated cases uses almost 100% of recycled paper and board – but the strength of this product depends on a proportion of new kraft fibre, which comes from abroad.

Newsprint – contains the weakest fibres because a newspaper has a short life span. Over the last few years, in discussions between newspaper publishers, newsprint manufacturers and the Government, agreements have been reached for UK newspapers to contain significantly greater amounts of recycled content year upon year. Initially, targets were achieved four years early, primarily

Figure 3.60 *Recovering energy from waste.*

Labels in figure: Steam turbine; Heat is used to make electricity; Gases are filtered and cleaned; Waste is incinerated; Waste is dumped in incinerator because of a new reprocessing plant in Kent. Future targets may prove challenging because of the growth in consumption of newsprint and the economics of investing in new reprocessing plants as opposed to using imported virgin fibres.

Although the recycling of paper and board may make economic and environmental sense, waste paper cannot be used in all paper grades or be used indefinitely. Three factors affect the use of recovered paper waste.

- **Strength** – every time a fibre is recycled it loses some of its strength. For papermaking purposes fibres can only be reused about six times.

- **Quality** – some paper grades make little or no use of recycled fibres because they require certain qualities only provided by the use of new fibres.

- **Utility** – it is not possible to recover all types of paper. Some are put to permanent use in books, etc. and some are disposed of when flushing the toilet. Others are not recoverable because they are part of a laminate or have a glossy surface finish.

Waste paper, or recovered paper as it is now often known, is the most important raw material for the UK paper and board industry. Paper waste is collected by either the Local Authority or by a waste paper merchant, sorted, graded and sent to the paper mill for recycling. The paper mill uses a hydrapulper filled with water to make the paper waste into a 'slush' and large contaminants are removed. The paper slush is then filtered and screened a number of times to make it more suitable for papermaking. Depending upon the quality of paper being produced, quantities of virgin pulp may be added. Four broad categories can be used to identify grades of waste paper based upon criteria such as the type of material the waste contains and how it will be used in the recycling process.

Remarkable Pencils

Remarkable Pencils Ltd. has successfully managed to recycle plastic vending cups into pencils and other stationery. The plastic contained in one High Impact Polystyrene vending cup is enough to make one remarkable pencil. Some people may argue that plastic vending cups are themselves unnecessary but daily production rates of 20,000 pencils means that 20,000 vending cups are being saved from disposal on landfill sites every day. The company has developed a strong, contemporary brand identity with its products so that consumers are more liable to purchase an environmentally aware alternative to the traditional wooden pencil.

THINK ABOUT THIS!

Conduct a study on the amount of waste generated by your school or college. Where does the majority of it come from – photocopying, litter, school dinners? How could this waste be minimised? Just think that your school is just one of thousands of schools in the UK alone all producing as much waste as you!

Table 3.22 The four grades of waste paper used for recycling.

Paper grade	Characteristics	Sources	Recycled applications
Pulp substitute grades	Top-quality waste that can be used with little need for cleaning.	Unprinted trimmings and offcuts from printers and converters	Printing and writing papers
De-inking grades	Grades from which the ink is removed before the recycling begins.	Office waste, newspapers and magazines	Graphic and hygienic (tissue) papers, newsprint
Kraft grades	Long strong fibres that generally come from unbleached packaging materials.	Paper sacks	New packaging including corrugated cases
Lower grades	Consist of mixed papers that are uneconomic to sort due either to the small quantities of each type or the level of non-recyclable material being too high.	'Junk' mail	Middle layers of packaging papers and boards

4 Renewable and non-renewable sources of energy

Since the industrial revolution in the 18th century, the burning of fossil fuels on an increasingly massive scale for generating power has resulted in large emissions of carbon dioxide, a greenhouse gas that contributes to global warming. The concentration of carbon dioxide in the Earth's atmosphere is increasing, raising concerns that solar heat will be trapped and the average surface temperature of the Earth will rise. Scientists estimate that a rise in global temperatures by 1½ to 2 degrees Celsius could cause unimaginable devastation through flooding, changes in weather patterns, desertification and displacement of entire populations. Therefore, a more sustainable solution for the planet's future energy needs based upon economic and environmental implications needs to be considered.

The 'Sandals' versus the 'Nukes'

Since global warming has become high on the environmentalist agenda, 'green' groups are split on the issue of which is the more sustainable – renewable or non-renewable forms of energy. On one side is the conventional green lobby, which believes in weaving ourselves more deeply into the natural world through the use of renewable energy sources like wind, solar and biofuels. On the other side is the less conventional opinion that humans should not place more demands upon nature but instead use technology such as nuclear energy and carbon-scrubbed natural gas.

The ideal power source has to produce the largest amount of energy achievable at an affordable cost, with as little environmental pollution as possible. Renewable energy is still in its infancy and is currently quite expensive to set up, but with time and more widespread use could become economically viable. However, its supply is unreliable – for example, sun for solar cells and wind for wind turbines – and there are problems with energy storage. Nuclear energy, on the other hand, is now an extremely safe way of producing energy with very low carbon emissions. Many scientists argue that a combination of the two is the best way forward.

FACTFILE:

'Sandals' versus 'Nukes'

'Sandals'	The 'Sandals' believe in the back-to-nature approach where humanity should be woven more deeply into the natural world through the use of renewable energy sources.
'Nukes'	The 'Nukes' are less conventional green thinkers who believe that technology is the way forward, therefore reducing our dependency on the planet's natural resources.

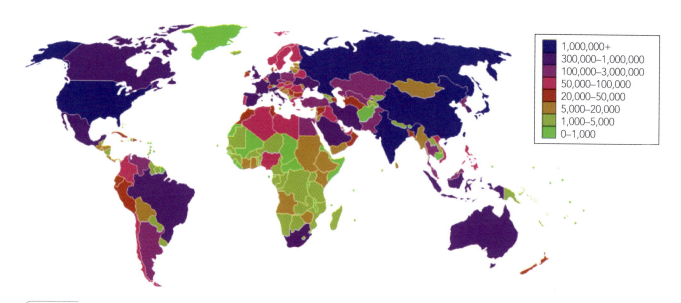

	1,000,000+
	300,000–1,000,000
	100,000–3,000,000
	50,000–100,000
	20,000–50,000
	5,000–20,000
	1,000–5,000
	0–1,000

Figure 3.61 *Carbon dioxide emissions by country.*

THINK ABOUT THIS!

Divide your class up into two groups: one will be the 'Sandals' and the other 'Nukes'. Research and hold a debate on the topic of renewable sources of energy versus nuclear energy. Record your findings.

You may also want to consult with your Geography department as this is also an issue for them – perhaps Geography students versus D&T students!

Table 3.23 Advantages and disadvantages of renewable sources of energy.

Energy source	Process	Advantages	Disadvantages
Wind	Power of wind turns turbines, which generate electricity.	• Flexibility – can be used in large-scale wind farms for national electrical grids as well as in small individual turbines for providing electricity to rural residences or grid-isolated locations. • Use of wind turbines is non-polluting, environmentally friendly and sustainable – produces more than 50 times as much energy over its lifetime as is consumed by its construction and installation. • Produces low-cost power if developed commercially, involving low marginal costs to run as fuel costs close to zero with relatively low maintenance costs. • Could be installed off-shore to minimise visual impact and take advantage of the moderate yet constant breezes.	• Can only provide a small proportion of total energy needs due to the amount of turbines needed in relation to space available for wind farms. • Unsightly on-shore wind turbines and wind farms spoil picturesque landscapes. • Infrastructure required for wind farms damages landscape. • Controversial – noise and vibration of moving turbine has potential to affect local community. • Affects environmentally sensitive coastal sites e.g. those with substantial bird life.
Water	Running water turns turbines and generates hydroelectric power (HEP)	• Fuel is not required so eliminates fuel costs and production of carbon dioxide. • Hydroelectric plants are highly efficient with minimum running costs due to highly automated operation. • Hydroelectric plants tend to have longer economic lives than fuel-fired generation, with some plants now in service having been built 50–100 years ago. • High initial set-up and construction costs recovered after only a few years due to sale of vast amount of electricity generated. • Reservoirs created provide improved leisure and tourism, e.g. water sports, fishing, etc. • Large dams can control flooding and protect towns downstream.	• Extremely expensive to construct dams and power plants. • Flooding of vast areas of land to create reservoir requires local population to be relocated. • Rivers may be diverted, which causes problems to local communities who rely on river for living. • Dam failures (accidental or sabotage) could cause massive destruction due to flooding. • Greenhouse gases produced can be high in tropical regions due to decay of plant life in reservoirs, producing methane. • Hydroelectric projects can be disruptive to surrounding aquatic ecosystems, e.g. affecting fish breeding and therefore birdlife. • Causes changes in the downstream river environment, e.g. erosion of river banks and low dissolved oxygen content of water.

Solar	Hot water and/or electricity generated from solar energy via solar panels and photovoltaic cells.	• Huge amounts of energy available from the sun. • Pollution-free during use. • Low operating costs and very little maintenance required after initial set-up. • Economically competitive especially for isolated or remote regions. • Produces enough electricity for the national grid to cope with peak demand times. • Local grid-connected solar electricity systems can be self-sufficient.	• Relatively expensive set-up costs for domestic and commercial buildings. • Currently, solar electricity can be more expensive than electricity generated by other sources. • Solar heat and electricity are not available at night and may be unavailable due to weather conditions, so a storage or complementary power system is required. • Energy lost by converting DC current generated into AC current for use in the national grid.
Biomass and biofuels	Plant materials are either incinerated to produce heat and electricity or biogas is produced from anaerobic digestion.	• Relatively inexpensive source of energy. • Large amounts of waste biomass materials available from agricultural processing and landfill. • Production of biogas reduces the release of methane into atmosphere – a harmful greenhouse gas. • By-products of biogas production can be sold and used as compost and fertiliser to improve soil condition.	• Ecological damage, including deforestation and intensive farming practices. • Currently, expensive processing costs of converting biomass into fuels with low yield. • Incineration causes carbon dioxide pollution.

Table 3.24 *Advantages and disadvantages of non-renewable sources of energy.*

Energy source	Process	Advantages	Disadvantages
Nuclear	A controlled nuclear chain reaction creates heat, which is used to boil water, produce steam, and drive a steam turbine which in turn generates electricity.	• Uses uranium, which is an abundant and widely distributed fuel. • Controlled chain reaction creates heat that can also be used to heat the power station. • Mitigates the greenhouse effect if used to replace fossil-fuel-derived electricity. • Passively-safe nuclear reactors use new technology leading to increased levels of safety to avoid leaks and overheating leading to meltdown. • Future development of fission reactors, which are cleaner and more efficient.	• Unpopular/mistrust with public due to media coverage of large-scale accidents, e.g. Chernobyl in 1986. • Problem of storing radioactive waste for indefinite periods, e.g. thousands of years to decay. • Potential for severe radioactive contamination by accident or sabotage and proliferation of nuclear weapons in some countries. • Mining of uranium causes damage to environment and pollution.
Fossil Fuels	The burning of hydro-carbons (oil, coal and gas) to produce heat and power.	• Economies of scale – large amount of electricity produced leading to low-cost energy supply. • Gas-fired power stations are very efficient. • Power stations can be built almost anywhere, including dedicated transport networks assuring large quantities of fossil fuels.	• Finite resources – coal, oil and gas will run out. • Largest source of emissions of carbon dioxide, contributing to global warming. • Generates sulphuric, carbonic and nitric acids, which cause acid rain. • Fossil fuels contain radioactive materials that are also released into the atmosphere. • Burning coal generates large amounts of fly ash. • Mining of coal and extraction of oil and gas cause damage to the environment and pollution. • Regional and global conflicts triggered over oil reserves.

WEBLINKS:

www.carbontrust.co.uk – The Carbon Trust helps business and the public sector cut carbon emissions, and supports the development of low-carbon technologies.

5 Responsibilities of developed countries

Global sustainable development

The challenge for people in developed countries (Great Britain, USA, etc.) is the need to reduce their use of scarce resources and reduce pollution, which is associated with over-consumption. This is a shift towards sustainable consumption and the reduction of an individual's 'carbon footprint'.

The challenges for developing countries (India, China, etc.) are different. In many instances people in developing countries need to consume more; for example, to gain greater access to clean water, electricity and health care. One method might be to trade more with developed countries to bring in much-needed foreign investment to domestic economies. Developing countries will need to have access to markets in developed countries in order to expand. However, developed countries need to shrink their markets to address over-consumption, which creates tighter and more impenetrable markets for developing countries to sell to.

The United Nations General Assembly authorises Earth Summits where representatives of all nations meet to discuss sustainable development. Most countries have established, with the World Summit on Sustainable Development (WSSD), some form of focal point or mechanism at the national level to oversee the implementation of the Earth Summit agreements, e.g. global trade or reduction in greenhouse emissions. All countries are invited to speak at these summits. Norway, for instance, has stated several practical steps towards sustainable consumption that would include:

- improving analysis, public awareness and participation

- providing incentives for sustainable consumption

- energy: sustainable use, efficiency and renewable sources

- implementing new strategies for transportation and sustainable cities

- accelerating the use of more efficient and cleaner technologies

- strengthening international action and cooperation.

Extract from the Report of the Symposium: Sustainable Consumption, Ministry of Environment, Norway, 1994.

Therefore, it follows that if global sustainable development is to succeed then all nations must firstly agree the terms and conditions and secondly, and more importantly, implement the changes needed.

THINK ABOUT THIS!

Organise a mini Earth Summit by dividing the class into two groups representing either developed (UK or USA) or developing (India or China) countries. What are the main issues for global sustainable development in each type of country? How can you agree upon resolutions to tackle these problems?

Impact of industrialisation on global warming and climate change

Kyoto Protocol

The Kyoto Protocol is an amendment to the United Nations Framework Convention on Climate Change, an international treaty on the contribution of human activities to global warming. The protocol sets targets for the reduction of greenhouse gas emissions (carbon dioxide and others) by 5 per cent of 1990 levels by the nations signed up to the agreement by 2012. The objective of the protocol is to stabilise greenhouse gas concentrations in the atmosphere at a level that would prevent dangerous changes to the world's climate. The treaty was negotiated in Kyoto, Japan in 1997 and came into force in 2005 following confirmation by Russia. As of December 2006, a total of 169 countries have signed up to the agreement, representing over 61.6 per cent of emissions from industrialised developed countries. Participating countries agreed on a set of "common but differentiated responsibilities".

- The largest share of historical and current global emissions of greenhouse gases has originated in developed countries.

- Per capita emissions in developing countries are still relatively low.

153

- The share of global emissions originating in developing countries will grow to meet their social and development needs.

The United Nations Framework Convention on Climate Change, 2006.

This means that developing countries are exempt from emission reduction targets in the Kyoto Protocol because they were deemed not to be the main contributors. Although developing countries such as China and India are undergoing rapid industrialisation and are exempt at present, they also share the common responsibility that all countries have in reducing emissions.

A notable exception to the Kyoto protocol is the United States which, although supporting it in principle, has never confirmed its participation. This may be partly due to the massive amounts of energy required to sustain its economy and a releuctance to accept targets imposed upon it. Instead, it has signed up to an agreement between Asia–Pacific nations including Australia, China, India, Japan, South Korea, and the United States. This Asia–Pacific Partnership on Clean Development and Climate agrees to cut emissions by developing cleaner energy supplies and technologies but without the enforcement of specific targets.

Non-Fossil Fuel Obligation

The Non-Fossil Fuel Obligation (NFFO) was instigated in 1989 when electricity generation in the UK was privatised. Originally, money raised by the associated Fossil Fuel Levy was used to subsidise UK nuclear power generators, which continued to be state owned. However, its scope was enlarged to include the renewable energy sector in order to offer financial support for renewable technologies.

The UK Government has stated that it wishes to halve its carbon emissions by 2050, which is an ambitious task. Therefore, the government superseded the NFFO with the Renewables Obligation in April 2002 stating that all electricity suppliers should source 10 per cent of their supply from renewable technologies by 2010, rising to 15 per cent by 2015. The most promising sources of alternative energy in the UK are wind, wave and tidal.

The development of wind energy is a rapidly expanding business and the UK has the largest potential wind energy resource in Europe. It is set to account for 8 per cent of electricity generation by 2010 and is capable of producing electricity for the National Grid at prices much lower than that of coal and nuclear. The cost of wind power has reduced considerably over the last few years due to the fall in cost of turbines, the increase in size of turbines meaning that fewer are needed and decreasing project development costs as developers have gained experience. Although wind farms have been controversial when located onshore, the recent trend for offshore locations could be a compromise that will benefit the whole of the UK.

Marine renewable technologies such as wave and tidal energies are a huge untapped resource for the UK, having the best wave and tidal resource in Europe. It has the

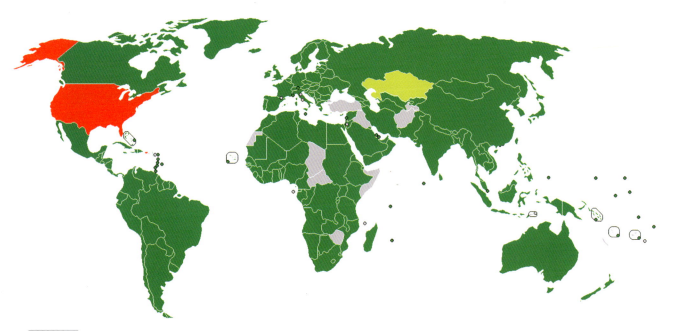

Figure 3.62 *Countries participating in the Kyoto Protocol.*

nPower Renewables © Anthony Upton 2003

Figure 3.63 *More-favourable offshore wind farms are set to make significant contributions to the UK's renewable energy production.*

potential of providing a considerable proportion of the UK's energy needs and a number of innovative marine energy devices are currently under development. However, such technologies face a number of challenges before they can become fully operational on a large-scale commercial basis.

THINK ABOUT THIS!

The development of offshore wind farms and wave and tidal technologies will affect the coastal regions of the UK. What effect could they have upon shipping and other commercial practices on which the UK relies?

WEBLINKS:

www.bwea.com – The British Wind Energy Association

www.climatechallenge.gov.uk – Understanding climate change in the UK

Reducing your 'carbon footprint'

A carbon footprint is a measure of the impact human activities have on the environment in terms of the amount of greenhouse gases produced, measured in units of carbon dioxide. These greenhouse gases (primarily carbon dioxide, methane and nitrous oxide) are a result of industrialisation and modern living and are contributing towards global warming. Every time you watch television, for example, you are producing carbon emissions because of the burning of fossil fuels in the generation of electricity. Therefore, it is everyone's responsibility to reduce their individual carbon footprint by firstly recognising how you personally impact global warming, including:

- annual household energy use, e.g. electricity and gas use
- annual travel i.e. car and public transport, flights, etc.

There are many simple ways in which an individual can save energy in the household, from installing energy-saving lightbulbs to turning electrical appliances off when not in use instead of leaving on 'standby'. In terms of travel, car pooling, public transport, cycling or walking are viable alternatives to using the car. However, when there are very few alternatives, such as a longhaul flight, carbon offsetting might be the answer.

Carbon offsetting is way of compensating for the emissions produced with an equivalent carbon dioxide saving. These can range in scale from planting trees in the UK to conservation of wildlife habitats in Africa or South America, as trees consume carbon dioxide and give out oxygen; these areas are known as 'carbon sinks'. Some of the larger projects can also benefit local communities by providing employment, which reduces poverty.

For manufacturers, their carbon footprint can be efficiently and effectively reduced by:

- applying life-cycle assessment (LCA) techniques to products in order to accurately determine the current carbon footprint

- identifying 'hot spots' in production processes in terms of energy consumption and associated carbon dioxide emissions

- optimising energy efficiency, so reducing carbon dioxide emissions and other greenhouse gases contributed from production processes

- identifying carbon offsetting solutions to neutralise the cabon dioxide emissions that cannot be eliminated by energy-saving measures.

THINK ABOUT THIS!

Use the carbon footprint calculator at www.carbonfootprint.com to work out just how much you impact global warming. What can you personally do to reduce your carbon footprint? Are there ways in which you can offset your carbon use?

WEBLINKS:

www.carbonfootprint.com

www.click4carbon.com – both contain lots of information on reducing your 'carbon footprint'

Sustainable timber production

The UK, for example, relies heavily upon the import of forest products and accounts for 8 per cent of global trade in tropical hardwoods. The developing countries that produce this timber benefit little from this trade, with only 10.5 per cent of the revenue from timber production benefiting the producing country. Timber has been the focus of considerable efforts over the past decade to establish more sustainable production and trading systems. There are a number of problems associated with forests.

- **Deforestation** – the full-scale removal of forest to make way for other land uses such as settlement, infrastructure and mining. Global deforestation is currently taking place at a rate of approximately 17 million hectares each year. Deforestation occurs because trees are being cut

at rates faster than forest regeneration and because many forests are being cleared to make way for other land uses. When reforestation occurs it is not of equal quantity or quality – it does not replace all the benefits of the natural forest.

- **Environmental degradation of forest areas** – linked to deforestation can be soil erosion, watershed destabilisation and micro-climate change. Industrial air pollution also reduces forest health.

- **Loss of biodiversity** – deforestation and environmental degradation contribute to a rapid reduction in ecosystem, species and genetic diversity in both natural and planted forests. Forest abuse in biodiverse tropical regions is of major concern, with some scientists estimating that 1 per cent of all species are being lost each year.

- **Loss of cultural assets and knowledge** – for indigenous peoples whose lives are destroyed by deforestation. Undocumented knowledge of nurturing the forest evolving through long periods of time is diminishing as forest area reduces.

- **Loss of livelihood** – for forest-dependent peoples, particularly in poor countries. Further social and economic problems are created elsewhere, such as in cities due to the redistribution of the local population.

- **Climate change** – both regional and global and contributes to global warming. Forests play a major role in carbon storage and with their removal more carbon dioxide enters the atmosphere, which may unbalance the Earth's carbon cycle.

Countries dependent upon the import of timber, such as the UK, have a responsibility to encourage the development of sustainable production and trading systems to minimise the amount of deforestation and its effects upon the environment, including:

- no longer importing from sources that involve deforestation

- moving to supply sources in areas of ecological surplus, e.g. the high-yield plantations of Brazil, Chile and New Zealand

- certification systems that ensure that forests producing goods for the UK are sustainably managed

- timber tracing systems to ensure that products from certified forests can be identified as such

- reducing consumption through education and advisory approaches that show how to produce the same benefits from less timber

- encouraging exporting countries to make the necessary policy changes required for the transition to sustainable forest management

- supporting international efforts to control the trade in unsustainably produced wood

- improving the aid process to poor communities involved in current deforestation methods.

WEBLINKS:

www.fsc.org – The Forest Stewardship Council (FSC)

Figure 3.64 *The Forest Stewardship Council (FSC) promotes environmentally appropriate, socially beneficial and economically viable management of the world's forests.*

Exam Café

Relax and prepare

By now you should have a sound knowledge and understanding of a range of key topics developed over your AS level studies. You will need to build on this and develop even more in-depth knowledge and understanding of some very modern D&T issues. For example, sustainability is not simply a topic for you to revise at A2 level but a very real global issue that will affect future generations.

Again, it is important to note that the questions asked by the examiner in this exam paper will cover aspects from all four sections of this unit. No paper will ever focus upon one section entirely. Therefore, it is vital that you have a secure knowledge and understanding across all four sections.

Revision Summary

You should give yourself plenty of opportunities to answer examination-style questions throughout the course so you are prepared for the final examination. Use the sample assessment materials (SAMs) and past exam papers provided by Edexcel (www.edexcel.org. uk) and the questions in this textbook.

Don't forget – if in doubt, ASK! Your teacher is there to help you understand the theory in this unit. Have a good, long think about appropriate questions to ask your teacher – it may be a good idea to discuss it with your peers first to see if they can explain it more clearly.

And finally, keep a set of well ordered and legible revision notes, which will help you to learn key topics and which you can always refer back to when in doubt.

Refresh your memory

Revision checklist

▷ Make sure that you have answered all the questions at the end of this section.

▷ Make sure that your revision notes are well ordered, clear and up-to-date.

▷ Use the web links to read around each key topic so that you are well informed.

▷ Use sample assessment materials and past papers to practice your exam technique.

▷ Discuss any problems with your peers or teacher – don't keep them to yourself!

Get the result!

Tips for answering questions

Questions at A2 level will be more difficult than those at AS level so it is extremely important that you read each question carefully before you respond. It might be a good idea to use a scrap piece of paper to outline your response if you think you have enough time.

Always look at the amount of marks awarded for each question in brackets. This will give you a good indication of how many points need to be raised in your response. As a general rule of thumb, look at the following command words and what you have to do in order to gain the marks:

Give, State, Name	(1 mark)	These types of questions will not feature heavily at A2 level but may appear at the beginning of the paper or question part. They are designed to ease you into the question with a simple statement or short phrase.
Describe, Outline	(2+ marks)	These types of questions ask you to simply describe something in detail. Some questions may also ask you to use notes and sketches; you can gain marks with the use of a clearly labelled sketch.
Explain, Justify	(2+ marks)	These types of questions will be commonplace in this exam. They are asking you to respond in detail to the question – no short phrases will be acceptable here. Instead, you will have to make a valid point and justify it.
Assess, Consider, Discuss	(3+ marks)	These types of questions require a more detailed response. You need to structure your own response, addressing a number of key points (possibly advantages and disadvantages) in your answer, sometimes coming up with a resulting argument or conclusion.
Compare	(4+ marks)	These types of questions will normally give you two different products and ask you to compare them either in terms of their design and manufacture or issues relating to sustainability.
Evaluate	(4+ marks)	These types of questions will appear towards the end of the paper or question part and are designed to stretch and challenge the more able student. They require you to make a well-balanced argument, usually involving both advantages and disadvantages.

Model answers

The following four questions should demonstrate the style of questions using some of the different types of command words. The places where marks have been awarded are indicated in brackets. These are referred to as 'trigger points' and are parts of the examiner's mark scheme where marks are expected to be awarded.

Exam question 1

Explain **two** advantages to the **manufacturer** of using an electronic point of sale (EPOS) system to gather sales information.

(4)

Martin

1. *It is faster at collecting sales information.* **(1)**

2. *It uses a barcode to identify the product.*

(1 mark)

Two statements are offered, neither of which are justified. The first statement is relevant, as faster collection of sales information would benefit the manufacturer, but why? The second statement simply notifies us of a component of EPOS and not an advantage.

David

1. *EPOS provides an easier and faster way of collecting sales information,* **(1)** *which can be used by the manufacturer to respond quickly to consumer demand.* **(1)**

2. *EPOS enables the manufacturer to operate a just-in-time system,* **(1)** *which reduces the need for large stock levels and so reduces costs.* **(1)**

(4 marks)

Two fully justified responses are given that clearly demonstrate the candidate's knowledge and understanding of EPOS when applied to the **manufacturer**. The focus of this type of question could quite easily be on the retailer or the consumer so it is really important that you read the question carefully.

Exam question 2

Discuss the benefits of using genetic engineering in the production of paper and board. (3)

Gary

Genetic engineering benefits the production of paper and board because it produces quicker-growing trees that are more resistant to disease **(1)** *so there is a plentiful supply of raw materials to make paper and board.* **(1)**

(2 marks)

The first part of the response contains two relevant points: 'quicker-growing trees' and 'more resistant to disease', but they are not justified. The second part of the response seems to pull the whole response together by making reference to there being trees grown specifically to make paper and board.

Charonne

Genetic engineering can be used to produce trees with reduced lignin content. **(1)** *This means that fewer chemicals need to be used to break down lignin in the chemical pulping process,* **(1)** *so it is kinder to the environment. The growing of genetically modified trees also involves better forest management, which reduces the problems of deforestation* **(1)** *caused by the vast amount of raw materials needed for the paper and board industry.*

(3 marks)

A good response that looks at the question from a sustainable point of view. The candidate makes the connection between the need for vast amounts of raw materials and how genetic modification of trees can be used in a positive way to supply the needs of the paper and board industry.

Exam question 3

Aesthetic design movements throughout history have influenced the styling of products and architecture. Outline the 'style' of **one** of the following aesthetic design movements:

- Art Nouveau
- Art Deco

(6)

Lesley

Art Nouveau was a movement that used flowing lines in its designs as opposed to Art Deco, which used zig-zag shapes. It was a time when people had more money and could afford to buy more products. They wanted things that looked good so they liked the idea of more stylised decoration such as flowing lines that looked like plants. (1) Women featured heavily in their designs with long, flowing hair. (1) Designers were also inspired by Japanese culture.

(2 marks)

You will see that the first sentence is not awarded a mark as the question asks students to outline the style of **one** movement only and not to compare two. The second sentence is simply waffle. The following two sentences gain marks because they described two styling points whilst the last sentence simply makes a statement with no explanation.

Jon

The style of the Art Nouveau movement is probably best characterised by the languid line (1). Designers found inspiration in natural forms (1) and represented them with curvy 'whiplash' lines and stylised flowers. (1) The peacock feather motif was often featured because it symbolised the hedonistic views of the time. (1) Art Nouveau was often referred to as 'feminine art' due to its frequent use of languid female figures with long, flowing hair. (1) The grid structures of Japanese interiors provided vertical lines and height to many pieces of furniture, (1) especially those of Charles Rennie Mackintosh. (6 marks)

When you read this response you can instantly tell that this student is well informed on the subject of Art Nouveau and has studied it in some detail. Again, the marks in brackets indicate where the marks have been awarded. Here, there are clearly six different aspects to Art Nouveau styling succinctly presented.

Exam question 4

Evaluate the use of fully automated production and assembly lines incorporating robots when manufacturing products compared with labour-intensive methods. **(10)**

Milton

Machines can do things quicker, better and more efficiently than humans. A robot can work for 24 hours a day without a break or getting tired but a human can't. (1) It also costs a lot to employ many workers when you don't have to pay a machine anything except that it costs a lot to buy in the first place. (1) People that work on production lines often become bored with their jobs as they do the same thing every day and they are usually not very well paid. (1)

(3 marks)

The candidate has started the response with a very basic statement that does not gain any marks at A2 level. The response focuses upon the idea that machines are more efficient than humans but fails to justify many of the points raised sufficiently. The candidate would have benefited perhaps from structuring the response in rough first in order to fully evaluate the topic.

Parul

Advantages of fully automated production over labour-intensive methods:

- *Automation increases productivity and reduces running costs (1) due to the efficiency of machines and less wages paid to a large manual workforce. (1)*

- *Robots can work freely in hazardous conditions such as paint shops, (1) which humans could not do without risks to health and safety. (1)*

- *Automation can free up the workforce from repetitive manual labour, (1) allowing more people to enter higher-skilled jobs, which are typically higher paying. (1)*

Disadvantages of fully automated production over labour-intensive methods:

- *Automated processes with robots are not suitable for extremely detailed manufacturing and finishing activities (1) where the human senses such as vision, touch and pattern recognition are still better. (1)*

- *Robots do not have the ability to learn and make decisions when the required data does not exist (1) whereas a human workforce can react quickly to change and adapt accordingly. (1)*

(10 marks)

Here, the candidate has structured the response in a logical manner as 'evaluate' questions ask for both advantages and disadvantages. The bullet-pointed responses under each heading are fully justified and demonstrate a very good understanding of the impact of automation upon manufacturing and employment. Note that more responses are given for advantages than disadvantages – this is perfectly ok as long as both are present.

Practice questions

1. Explain **two** advantages of using virtual modelling and testing in the development of a new product. **(4)**

2. Explain **three** ways in which computer-integrated manufacturing (CIM) benefits a manufacturer mass-producing a product. **(6)**

3. Discuss the impact of the development of industrial mass-production on:

   (i) workers. **(3)**

   (ii) consumers. **(3)**

4. Assess the effects of using biodegradable polymers, such as Biopol®, on the packaging industry. **(6)**

5. Outline the use of **one** smart material for an innovative application. **(4)**

6. Discuss the relationship between ergonomics and anthropometrics. **(4)**

7. Consider the ergonomic factors involved when designing a bicycle for an adult user. **(6)**

8. Figures 1 and 2 below show two different lemon squeezers.

   The lemon squeezer in Figure 1 is called 'Juicy Salif', designed by Phillipe Starck, and Figure 2 shows a widely available lemon squeezer from a high-street shop.

Figure 1

Figure 2

   Compare the lemon squeezer in Figure 1 with the lemon squeezer in Figure 2 with reference to the following:

   (i) form.

   (ii) function. **(8)**

9. Evaluate the use of alternatives to fossil fuels in the transportation of products from the manufacturer to the retailer. **(8)**

10. Evaluate the impact of recycling upon sustainable product design. **(8)**

Unit 4:

Commercial Design

Summary of expectations

1 What to expect

In this unit, you are given the opportunity to apply the skills you have acquired and developed throughout this course of study, and to design and make a product of your choice that complies with the requirements of a Graphic Products project. You are encouraged to be creative and adventurous in your work. Throughout this project you are expected to take ownership of all aspects of your work, and to take total control of your responses with your teacher as facilitator. You are strongly advised to target assessment criteria effectively in order to maximise your achievements.

In order to reach high attainment levels, you must adopt a commercial design approach to your work, reflecting how a professional designer might deal with a design problem and its resolution. The choice of design problem should have a real commercial use, in that it should be useful to a wider range of users and not simply be designing for yourself. The product should be designed to be commercially manufactured, i.e. batch- or mass-produced, unless it has been specifically commissioned as a 'one-off'.

The design problem should provide opportunities for a client or user group to have input into decision-making at various stages of the design and make process. A client or user group is defined as any third party you identify that is referred to and who can give informed critical feedback at various stages throughout the design process. Clients and user groups do not need to be specialists or experts; they can be drawn from any relevant group of people and may include other students, friends or family members.

A key feature of this unit is for you to consider issues related to sustainability and the impact your product may have on the environment. You may choose to design and make a sustainable product but, if you do not, you should still consider the issues of sustainability at relevant points in your designing and making activities. Sustainable issues could include materials production and selection, manufacturing processes, use of the product and its disposal/recycling.

This unit is set and marked by your teachers, then sent to Edexcel for moderation (sampling and checking of teachers' marks).

2 What is a Graphic Products project?

Graphic Products projects have two clearly defined pathways, either 'conceptual design' or 'the built environment'.

(i) **Conceptual design** projects incorporate a wide range of 3D products with associated graphics, for example:
- packaging design
- product/industrial design
- point-of-sale display
- vehicle design.

A range of modelling materials, including resistant materials (not a compulsory requirement), can be used; for example, the use of Styrofoam™ and/or medium-density fibreboard (MDF) for concept modelling.

(ii) **The built environment** project focuses on the surroundings that provide the setting for human activity, for example:
- architecture
- interior design
- exhibition design
- theatre sets
- garden design.

Built environment projects must contain a 2D and 3D element. However, the main emphasis must be on the 3D element, with the 2D element focusing on presentation graphics and technical drawings. A range of modelling materials, including resistant materials (not a compulsory requirement), can be used; for example, the use of foam board, polymers and wood for architectural modelling.

3 How will it be assessed?

The coursework requirement at A2 Level is a full design and make activity, offering you the opportunity to demonstrate the knowledge, skills and competencies that you have gained from your AS studies. The assessment criteria statements are broken down in the To be successful you will sections of this textbook.

Where large numbers of marks are assigned to assessment sections such as design and development and making, these have been broken down into smaller sub-sections to allow clearer and easier access to marks.

The maximum amount of marks available for this unit is 90.

Sections	Sub-sections		Marks
Product design and make	A. Research and analysis		4
	B. Product specification		6
	C. Design and development:	Design	10
		Review	4
		Develop	10
		Communicate	6
	D. Planning		6
	E. Making:	Use of tools and equipment	9
		Quality	16
		Complexity/level of demand	9
	F. Testing and evaluating		10
	Total marks:		90

Please note: It is extremely important that you sign the authentication statement in your Candidate Assessment Booklet (CAB) before your work is marked. If you do not authenticate your work Edexcel will give you zero credit for this unit.

4 The coursework project folder

This unit results in the development of an appropriate product supported by a design folder. The folder, which should include information and communication technology (ICT)-generated

images where appropriate, can only be submitted on A3 paper and is likely to be no more than 30 pages long. You can also submit your work electronically for moderation provided it is saved in a format that can be easily opened and read on any computer system, i.e. a PDF document.

Your product design and make folder must be organised in a clear and logical manner that reflects the order of the assessment sections. It is important that each page is evidenced in the appropriate section. This will allow your teacher to easily mark your work and provide your Edexcel moderator with a clear indication of your skills and ability.

Suggested contents		Suggested page breakdown
Title page with specification name and number, candidate name and number, centre name and number, title of project and year of submission.		Extra page
Contents page		Extra page
A. Research and analysis		3–4
B. Product specification		1–2
C. Design and development:	Design	3–4
	Review	1–2
	Develop	3–4
	Communicate	(Evidenced throughout section) Working drawings 1–2 Pictorial drawings 1–2
D. Planning		2–3
E. Making:	Use of tools and equipment	3–4
	Quality	
	Complexity/level of demand	
F. Testing and evaluating		2–3
Bibliography		Extra page
Total pages:		20–30

5 How much is it worth?

The product design and make coursework project is worth 60 per cent of the A2 course and 30 per cent of the overall full Advanced GCE.

Unit 4	Weighting
A2 level	60%
Full GCE	30%

Product design and make

Getting started!

In order to get this unit started, you must first identify a specific need or problem and derive from it a detailed design brief. Don't forget that in this unit you must adopt a commercial design approach, acting like a professional designer. Therefore, your choice of design problem should have a real commercial use and involve a real client or specific user group.

Table 4.1 Ideas for commercial design projects.

Problem/need	Client/user group	Outcome(s)
Your local youth club is run-down and underused and drastically needs refurbishing and updating.	Client: local community youth workers. User group: youth club members.	3D: interior model of new youth club. 2D: publicity materials for launch event.
Your local feeder-primary school needs an educational pack to teach pupils about green issues.	Client: teachers/ Head teacher at primary school. User group: primary school pupils.	3D: actual teaching aids and themed carry-case. 2D: worksheets and fact-sheets on green issues.
Your Dad's company is to exhibit at a national exhibition centre.	Client: your Dad and other members of his company.	3D: model of portable exhibition stand. 2D: banners and displays to attract attention to stand.

Design brief

To design and model a new Post-16 area at the front of The Ravensbourne School, which is currently used as a visitors' car park. At present, the gates where students walk into school are very unsafe due to the fact that both cars and pedestrians enter the school through the same narrow gate. The layout of the car park is very restricted and therefore could be utilised in a far more useful way to provide additional Post-16 facilities for the rapidly expanding sixth-form.

I intend to redesign this area so that it becomes an external Post-16 social/study area. Currently there is nowhere for sixth-formers to sit and work outside and I believe that providing an area where students can work outside would allow them to study and relax in more comfortable surroundings when the weather is fine. It is also conveniently located at the front of the school, which would act as an advertisement for Post-16 recruitment.

Clients:
- Mrs. Judah, Assistant Head
- Ms. Mills, Director of Post-16

User groups:
- Post-16 students (Year 12 and Year 13)

Figure 4.1 An example of an appropriate design brief.

ACTIVITY:

Take a long, hard look around your school or college site. Identify ten aspects of the site that you don't like, that don't work or are seriously underused. For example, where can students go to eat their lunch? Is this area safe and comfortable? Is it sheltered from the rain and sun? Are there enough spaces to sit at a table?

Produce a report entitled Ten things I don't like about my school that could be put right with good design. Compile the top ten in order of your dislikes and compare these with your classmates, with the number-one spot being something that potentially could be a good project.

There are ample opportunities for gaining client feedback throughout this project from site staff, teachers and even the Head teacher, as well as user-group feedback from the students themselves.

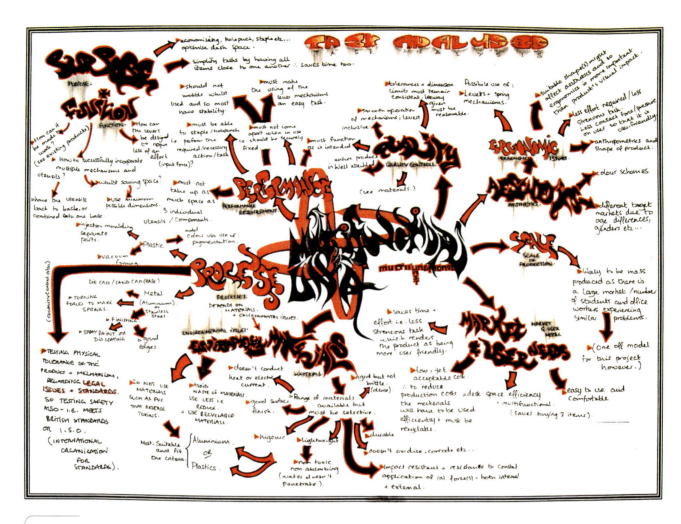

Figure 4.2 *This student has explored the task in detail by analysing the key words in the form of a spider diagram.*

A Research and analysis (4 marks)

Once you have identified an appropriate need and written a detailed design brief, you must analyse the need in order to focus on the research needed to help your work progress. You should use a range of research strategies to gather useful and relevant information that will help with your designing and making activities.

When gathering information, it is important that you are clear about what you need to find out. Research should be highly selective, ensuring that the information gathered is useful and relevant to the client/user group's needs identified and finalised during analysis. Research

should be focused and succinct and contain no worthless padding. You should avoid downloading large amounts of information from the Internet, or cutting and pasting from catalogues and databases, without providing detailed annotation to explain the selected information. All sources must be fully acknowledged to avoid risk of plagiarism.

A good starting point for research and analysis could include an interview or discussion with the client/user group to establish their thoughts and preferences regarding the proposed product. This information should be used to guide your analysis and research activities. In the analysis, you should ensure that you focus closely on the identified need, avoiding any general statements that are of no use and could be applied to any design situation.

Research could include the analysis of existing similar products to find out about materials, processes and construction methods used in commercial manufacture. Market research will allow you to test the viability of your intended product beyond the needs of the client/user group. Surveys or questionnaires should be designed carefully, avoiding questions that are general and useless in helping with the design process. A questionnaire should not be included simply for the sake of doing so; its use and the questions asked within it should be justified.

When researching materials, components and processes, you should take into consideration the concept of 'sustainability' so that you are able to make responsible and informed decisions about the impact of materials and resources upon the environment.

When all information gathering has been completed, you should analyse your research in order to help write a product specification that is relevant, meaningful and measurable.

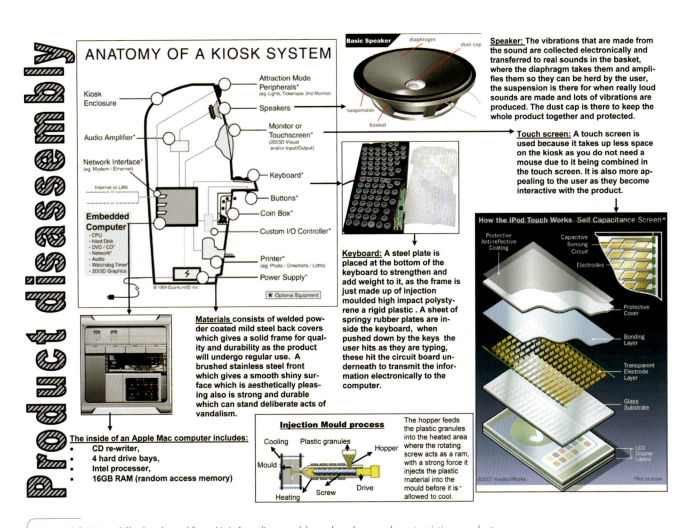

Figure 4.3 Using skills developed from Unit 1 to disassemble and analyse a relevant existing product.

To be successful you will:

A. Research and analysis

Level of response	Mark range
Produce detailed analysis with most design needs clarified. (1 mark) Present selective research that focuses on the needs identified in the analysis. (1 mark)	3–4
Produce limited analysis with some design needs clarified. (1 mark) Present superficial research that does not focus on the needs identified in the analysis. (1 mark)	1–2

B Product specification (6 marks)

It is important that you develop and write a detailed specification, as it will be used throughout the design process to review your ideas and their development and to check that the design requirements and client/user group needs are being satisfied. The specification should be used as a basis for testing and evaluating the completed product and any future modifications suggested should be referenced to specification criteria in order to check the success of your final product.

The starting point for a successful specification should be after the research and when essential requirements have been established as a result of studying the information gathered. You should consult with your client/user group to agree the specification points and to ensure that the criteria meet their needs. When specifying materials, components and processes, you should consider sustainability, and make decisions based on the environmental costs of extracting and processing the selected materials and the product manufacture, lifespan and disposal.

When writing a specification, you should try to avoid a rambling collection of points. The specification should be informed by your research findings. An effective specification is organised logically and could be achieved by using sub-headings such as:

- **purpose** – what is the aim or end-use of the product?
- **form** – what shape/style must the product take?
- **function** – what must the product specifically do?
- **user requirements** – what qualities must the product have to make it attractive to client/user group?
- **performance requirements** – what technical considerations need to be achieved within the product?
- **materials and components** – what materials and components should be used to aid performance?
- **size** – what physical dimensions are required?
- **safety** – what factors need to be considered to make the product safe to use?
- **quality** – how can a high-quality product be assured?
- **scale of production** – how many are to be made and by what manufacturing processes?
- **cost** – what are the considerations in determining cost?

Each specification point should contain more than a single piece of information, so that each statement is fully justified by giving a reason for the initial point. For example, it is not sufficient to say 'the material used is polystyrene', as this is not justified until 'because it is tough and can be injection moulded' is added.

Specification points should be technical and measurable where possible, so that testing and evaluation can be realistic. It is extremely important that your specification points are not superficial or general.

ACTIVITY:

Using the example of the bus shelter design specification in Figure 4.4, outline the tests that could be made to determine whether each point has been matched. For example:

- Be a sufficient length to accommodate five seated people and anthropometric height for the average user.

This is measurable as you could easily test how much space a seated person needs to occupy to be safe and comfortable. You can then obtain the minimum length of the bus shelter by multiplying this figure by five. Anthropometric data charts can be referred to in order to determine a height suitable for the majority of the population (95th percentile).

Design brief
To design and model a new bus shelter for Transport for London.

Client: Mr. Turner, Manager at 'First' (bus operator)

User group: Post-16 students at The Ravensbourne School

Product Specification
The bus shelter must:

Purpose:
- Provide a safe, comfortable and sheltered area in which to wait for buses on public transport routes operated by First.

Form:
- Have a modern style to represent upgrading on public transport.
- Incorporate strong Transport for London branding so that it is easily recognisable as a bus shelter.

Function:
- Provide shelter for at least ten people from the rain and shade from the sun.
- Provide seating for at least five people that must be comfortable to sit on but not allow people to lay across or sleep on.
- Provide travel details such as timetables, routes and expected arrival times for buses on individual routes.

User requirements:
- Indicate when a bus is going to arrive or warn people when a bus is going to be late.
- Provide sufficient shelter so that gusts of wind and rain cannot blow through or shading from direct sunlight whilst waiting for a bus.
- Provide full visibility to people inside and include lighting at night so people feel secure.

Performance requirements:
- Be securely anchored to the pavement so that it cannot be tipped over.
- Not include large flat surfaces, which could act as sails and cause the structure to stress in high winds.

Materials and components:
- Be manufactured using a stainless steel structure to prevent corrosion when exposed to British weather conditions.
- Make use of toughened safety glass, which is transparent but will not cause injury if vandalised.
- Use seating made from a suitable polymer or sheet metal that can be formed into interesting/comfortable shapes.
- Incorporate screens to display travel information that are encased and protected from vandalism.

Size:
- Not occupy more than half the width of a wide pavement so that pedestrians are free to walk by.
- Be of sufficient length to accommodate five seated people and anthropometric height for the average user.

Safety:
- Not contain any sharp corners/edges or low beams, which could cause injury to people waiting.
- Be relatively resistant to vandalism by ensuring that components cannot be readily removed or glass smashed into dangerous shards.

Quality:
- Use materials and components that have been subject to British Standard testing e.g. Kitemark on safety glass.
- Be manufactured using appropriate quality control (QC) procedures to assure a high-quality outcome.

Scale of production:
- Be able to be batch produced as approximately 100 are to be installed around the Bromley area.
- Be able to be manufactured using efficient fabrication techniques with use of standard components where necessary.

Cost:
- Come in on budget at £1,000 whilst still meeting the necessary requirements of the client and user group.

Figure 4.4 *A sufficiently detailed product specification that would enable a student to effectively review design ideas and test the final outcome.*

To be successful you will:

B. Product specification

Level of response	Mark range
Compile specification points that are realistic, technical and measurable. **(1 mark)** Produce a specification that fully justifies points developed from research in consultation with a client/user group. **(1 mark)** Realistically consider the sustainability of relevant resources when developing specification points. **(1 mark)**	4–6
Compile specification points that are realistic but not measurable. **(1 mark)** Produce some specification points that are developed from research in limited consultation with a client/user group, but are not justified. **(1 mark)** Superficially consider the sustainability of resources when developing specification points. **(1 mark)**	1–3

C Design and development

FACTFILE:

- Design and development carries large numbers of marks but is sub-divided into four areas – design, review, develop and communicate – which are explained individually to show what should be presented as evidence to gain marks.

- An important feature of this section is that you should consider issues related to sustainability and the impact your product might have on the environment.

Design (10 marks)

In this section, you have the opportunity to apply your design skills and the advanced knowledge of materials, components, processes and techniques developed through your experience of AS units. You should produce alternative design ideas that are realistic, workable, and that address the needs identified in your specification. Designs should be annotated and include as much detail as possible of materials, components, processes and techniques that could be used to construct each design idea.

A suitable starting point for design ideas might be to firstly explore shape and form. These aesthetic considerations are the most obvious visual reference by which you can begin to add more detail to your design ideas. For example, if you decide upon a suitable shape then you can begin to consider how it could be made and what materials and components are necessary. If you simply explore the aesthetics of your product or environment then you will not be able to access the higher marks. Remember: your ideas must be workable, so a sufficient degree of detail is necessary to communicate your intentions and test whether or not it can be developed further.

ACTIVITY:

Once you and your classmates have completed a series of design sheets for your project:

- Swap folders with the person next to you and study their designs in detail in order to carry out a critique of their work.

- Make a list of things that are not communicated as well as they could be, e.g. a sketch is unclear and requires additional annotation to explain a concept.

- Make a list of problems that you can see with some of their designs, e.g. the material chosen may not have sufficient properties for its application; suggest alternatives.

- Politely feed back your observations using constructive criticism and identifying areas for further improvement.

Once this activity is complete and everyone has had some feedback, the necessary amendments should be made to your project. An activity like this can have a positive effect upon your project as feedback is coming from your friends rather than from your teacher.

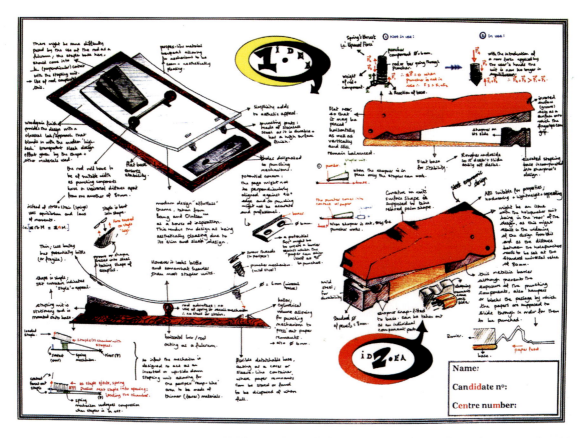

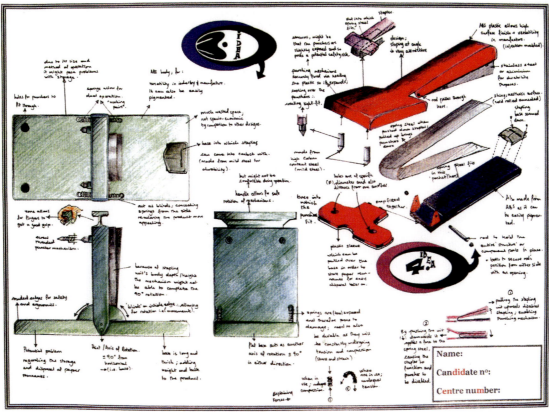

Figure 4.5 A student's highly visual design sheets explore ideas in detail.

Figure 4.5 A student's highly visual design sheets explore ideas in detail. (continued)

To be successful you will:

C. Design and development – **Design**

Level of response	Mark range
Present alternative ideas that are realistic, workable and detailed. **(1 mark)** Demonstrate a detailed understanding of materials, processes and techniques in ideas that is supported by research information. **(1 mark)** Address all specification points in ideas. **(1 mark)** Show client/user-group feedback. **(1 mark)**	7–10
Present alternative design ideas that are realistic and workable. **(1 mark)** Use relevant research to produce detailed ideas. **(1 mark)** Address most specification points in ideas. **(1 mark)**	4–6
Present alternative design ideas that are similar and simplistic. **(1 mark)** Use limited research to produce similar ideas. **(1 mark)** Address a limited number of specification points. **(1 mark)**	1–3

Review (4 marks)

An important part of your designing is to review and objectively evaluate your design ideas as they are produced. The comments made in reviewing design ideas should be based on objective, formative evaluation of each idea and should always be referenced to the specification in order to check the idea's potential in fulfilling your client/user group's need. It is always good practice to choose more than one idea for a detailed review as this gives the client/user group a choice of suitable outcomes.

Please note that you should not use simple 'tick-boxes' when reviewing design ideas, as this is always subjective and worthless in evaluating ideas against a specification effectively. In addition, avoid simple Yes/No answers as they do not allow any useful decisions to be made when deciding whether specification points have been met. **Remember – all specification points need to be justified.**

Do not feel that all of your specification points have to be successfully satisfied at this stage. No design is ever perfect first time around and that is why large amounts of money are used in the development of any product. Your initial design ideas should address the majority of the criteria but it is at the development stage that you will have to ensure that all specification points are met. At this review stage you are simply pointing out areas for further development.

As part of your review, design ideas should be discussed with your client/user group to ensure, through feedback, their suitability for their intended purpose. Information should be communicated through logical and well-organised statements, using specialist technical vocabulary. In addition, you should consider and justify some of their design decisions with reference to sustainability.

ACTIVITY:

Choose your best two design ideas, photocopy them and present them on a board for your client/user group to review. Prepare a short presentation of each idea outlining the concept behind the idea and be prepared to answer any questions that your client/user group may have.

Ask your client for their opinions on both designs – what they like/dislike and what they would like to see in any future designs. It might be a good idea to have prepared a brief questionnaire to prompt the discussion and record their views. Alternatively, conduct an interview using a Dictaphone to record the conversation so that you don't miss any relevant information.

To be successful you will:

C. Design and development – **Review**

Level of response	Mark range
Present objective evaluative comments against most specification points that consider client/user-group feedback. **(1 mark)** Include evaluative comments on realistic issues of sustainability relating to design and resources. **(1 mark)**	3–4
Present general and subjective comments against some specification points. **(1 mark)** Superficially evaluate an aspect of sustainability. **(1 mark)**	1–2

Review

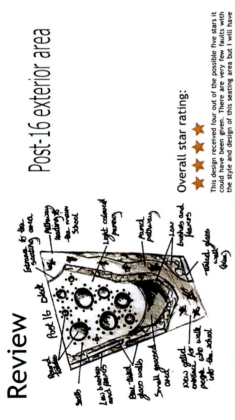

Post-16 exterior area

Overall star rating:

★★★★

This design received four out of the possible five stars it could have been given. There are very few faults with the style and design of this seating area but I will have

Specification

Function:

- There is a good seating area provided with this design that allows students to work, eat and do everything else they need to do during free periods, break and lunch times.
- It's easy to clean and maintain, e.g. Plants which don't need a lot of attention, furniture which is weather resistant and so on.
- There are calming colours and a relaxing layout with minimal clutter. Although there is a large amount of tables and chairs which may make the area uncomfortable to sit in when there is a large number of people using the seating area.
- There is an entrance which Post 16 students can get to the seating area via a door of the Post 16 café and there is a pathway leading from the road to the school and the seating area.
- A good litter system.
- The area has a modern and attractive look so that it both looks good and attracts people's attention to the school.
- The area is easy to maintain and not have a large impact on the environment around the school or look out of place compared to the rest of the school.
- There isn't anything that is likely to put anyone in danger or cause injury.

Market:

- The area will be attractive and somewhere Post 16 students would enjoy using.
- It will be somewhere that both 6th formers and teachers can comfortably use.

Aesthetics:

- It has a modern look using plants and materials which are both durable and attractive.
- There will be tables and chairs which are comfortable, fit in with the style of the area and which are long lasting.
- Plants will be used to make the area more calming and can also be used to block out the noise and view of the road.

Quality:

- Everything will be finished to a high standard. The paving must be laid perfectly so that it doesn't look messy and can't cause injury to anyone who may trip on it.
- All of the plants that will be used in the Post-16 are will be ones that can easily be maintained to a high standard without having a large amount of care.
- Everything in the seating area will be weather resistant so that it lasts a long time and doesn't become a safety hazard or unattractive.

Safety:

- All of the paving, plant pots and all the other things that will be in the seating area will all have been carefully placed and finished to a high standard so to not cause injury to anyone.
- Anything that may fall over will be secured so that it can't cause anyone harm.

Client Feedback

Miss Mills– Head of the Post-16

"This design looks very attractive and would definitely fit in with the modern Post-16 block. The round tables and blue glass tinted walls will all go well with the shape and colour of the building the seating area will be linked to. Making it automatically fit in and look right in the school.

I also really like how you have given this space a calming and tranquil look by using a large amount of plants. This will make the seating area a nice place to sit, eat and work in for both students and teachers.

Another aspect of this garden that I think that the school needed to have looked at is the problem of having pedestrians and cars using the same narrow entrance to the school. Incorporating a new entrance and pathway leading into the school is a far safer way for students to enter the school and will make it easier to monitor who enters and leaves the school.

Although I like this design there are a couple of things that I would like you to modify. I think that there needs to be a larger grassed area for students to use. Also I think there needs to be some form of feature in the garden that makes it stand out and look more attractive."

3rd Party Feedback

Post-16 students at The Ravensbourne School feedback of what they think about this design of the seating area:

Josh

"I like this design the most! I think it looks really modern and with the metal tables would look very attractive. I like how you have added a new entrance into the school and made the seating area look very spacious. If I was to pick a new design for the Post-16 seating area this would be it. Although I think there should be more small trees in the area."

Renée

"I really like this design for the seating area. I think that this would be a really good way to redesign the parking area outside of the Post-16 block. The area looks like it would be really calming and make the school look much better. Although I think should be something more in the area to make it more unique from anything any other school has."

Dan

"I think there needs to be a larger grassed area and fewer tables in this area. I don't think that many people would want to sit on the chairs but they would like to sit on the grass. I also think there needs to be a fountain or some form of water feature. Maybe the grass around the outside of the area needs to be thicker or a stronger material to make sure it doesn't get broken."

Nicola

"This area would definitely make the school look far better and give it more of an aesthetic appeal to visitors. I really like how you have added a new entrance for students to walk into the school and how you have kept it separate from the seating area, using the plants and bushes. I would use this area if it was made and I'm sure everyone else would."

Figure 4.6 A student's review of a design idea against the specification criteria with client and third-party feedback.

Develop (10 marks)

In this section, you will develop a final design proposal in consultation with your client/user group. Development of the final design proposal will give you the opportunity to bring together the best and most appropriate features of your initial design ideas. This refined final design proposal should not only meet all of the requirements of your product specification but also satisfy your client/user group needs.

You must show the development of your design, demonstrating how it has changed and moved on from initial ideas, using the results of review/evaluation and client feedback. It is not good practice to simply take an initial idea, make superficial or cosmetic changes, and then present it as a final developed proposal.

You should include as much detailed information on all aspects of the developed design as possible, including technical details of materials and components and their selection, processes and techniques. This is an opportunity for you to demonstrate an advanced knowledge and understanding of design and make activities.

Modelling should be used to test features such as proportions, scale, function, sub-systems, etc. Modelling can be achieved through the use of traditional materials, or 2D and/or 3D computer simulations. Evidence of modelling should be presented through clear, well-annotated photographs. Consultation with the client/user group should be evidenced in order to justify and clarify final design details.

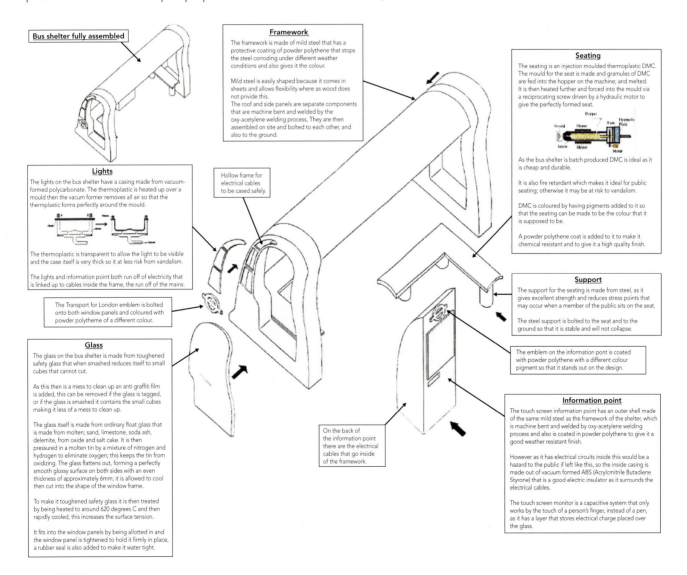

Bus shelter fully assembled

Framework

The framework is made of mild steel that has a protective coating of powder polythene that stops the steel corroding under different weather conditions and also gives it the colour.

Mild steel is easily shaped because it comes in sheets and allows flexibility where as wood does not privide this.
The roof and side panels are separate components that are machine bent and welded by the oxy-acetylene welding process, They are then assembled on site and bolted to each other, and also to the ground.

Seating

The seating is an injection moulded thermoplastic DMC. The mould for the seat is made and gramules of DMC are fed into the hopper on the machine; and melted. It is then heated further and forced into the mould via a reciprocating screw driven by a hydraulic motor to give the perfectly formed seat.

As the bus shelter is batch produced DMC is ideal as it is cheap and durable.

It is also fire retardant which makes it ideal for public seating; otherwise it may be at risk to vandalism.

DMC is coloured by having pigments added to it so that the seating can be made to be the colour that it is supposed to be.

A powder polythene coat is added to it to make it chemical resistant and to give it a high quality finish.

Lights

The lights on the bus shelter have a casing made from vacuum-formed polycarbonate. The thermoplastic is heated up over a mould then the vacuum former removes all air so that the thermplastic forms perfectly around the mould.

The thermoplastic is transparent to allow the light to be visible and the case itself is very thick so it at less risk from vandalism.

The lights and information point both run off of electricity that is linked up to cables inside the frame, the run off of the mains.

Hollow frame for electrical cables to be cased safely.

The Transport for London emblem is bolted onto both window panels and coloured with powder polythene of a different colour.

Support

The support for the seating is made from steel, as it gives excellent strength and reduces stress points that may occur when a member of the public sits on the seat.

The steel support is bolted to the seat and to the ground so that it is stable and will not collapse.

The emblem on the information pont is coated with powder polythene with a different colour pigment so that it stands out on the design.

Glass

The glass on the bus shelter is made from toughened safety glass that when smashed reduces itself to small cubes that cannot cut.

As this then is a mess to clean up an anti graffiti film is added, this can be removed if the glass is tagged, or if the glass is smashed it contains the small cubes making it less of a mess to clean up.

The glass itself is made from ordinary float glass that is made from molten; sand, limestone, soda ash, delemite, from oxide and salt cake. It is then pressured in a molten tin by a mixture of nitrogen and hydrogen to eliminate oxygen; this keeps the tin from oxidizing. The glass flattens out, forming a perfectly smooth glossy surface on both sides with an even thickness of approximately 6mm; it is allowed to cool then cut into the shape of the window frame.

To make it toughened safety glass it is then treated by being heated to around 620 degrees C and then rapidly cooled, this increases the surface tension.

It fits into the window panels by being allotted in and the window panel is tightened to hold it firmly in place, a rubber seal is also added to make it water tight.

On the back of the information point there are the electrical cables that go inside of the framework.

Information point

The touch screen information point has an outer shell made of the same mild steel as the framework of the shelter, which is machine bent and welded by oxy-acetylene welding process and also is coated in powder polythene to give it a good weather resistant finish.

However as it has electrical circuits inside this would be a hazard to the public if left like this, so the inside casing is made out of vacuum formed ABS (Acrylcmitrile Butadiene Styrone) that is a good electric insulator as it surrounds the electrical cables.

The touch screen monitor is a capacitive system that only works by the touch of a person's finger, instead of a pen, as it has a layer that stores electrical charge placed over the glass.

Figure 4.7 *An exploded drawing gives this student an opportunity to outline industrial and commercial manufacture when developing the product.*

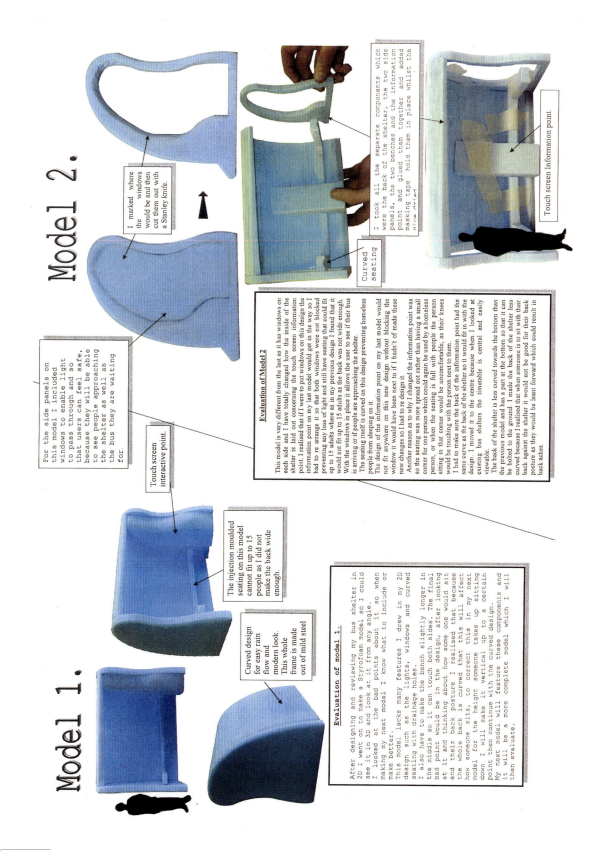

Model 2.

I marked where the windows would be and then cut them out with a Stanley knife.

I took all the separate components which were the back of the shelter, the two side panels, the two benches and the information point and glued them together and added masking tape hold them in place whilst the glue dried.

Touch screen information point.

Curved seating

For the side panels on this model I included windows to enable light to pass through and so that users can feel safe, because they will be able to see people approaching the shelter as well as the bus they are waiting for.

Touch screen interactive point.

Evaluation of Model 2

This model is very different from the last as it has windows on each side and I have totally changed how the inside of the shelter is laid out concerning the touch screen information point. I realised that if I were to put windows on this design the information point on my last model would get in the way so I had to re arrange it so that both windows were not blocked preventing any loss of light and still have seating that could fit up to 15 adults where as in my previous design I found that it would not fit up to 15 adults as the back was not wide enough. With the windows in place it allows the user to see if their bus is arriving or if people are approaching the shelter.

The seating itself is curved on this design preventing homeless people from sleeping on it.

The design of the information point on my last model would not fit anywhere on this new design without blocking the window it would have been next to if I hadn't of made these new changes so I had to re design it.

Another reason as to why I changed the information point was so the seating was more spread out rather than having a small corner for one person which could again be used by a homeless person, or when the seating is full with people the person sitting in that corner would be uncomfortable, as their knees would be touching with the person next to them.

I had to make sure the back of the information point had the same curve as the back of the shelter so it would fit in with the design. I moved it to the centre because when I looked at existing bus shelters the timetable is central and easily viewable.

The back of the shelter is less curved towards the bottom than the previous model and has a part at the bottom so that it can be bolted to the ground. I made the back of the shelter less curved because I realised that when someone is to sit with their back against the shelter it would not be good for their back posture as they would be leant forward which could result in back aches.

Model 1.

The injection moulded seating on this model cannot fit up to 15 people as I did not make the back wide enough.

Curved design for easy rain flow and modern look. This whole frame is made out of mild steel

Evaluation of model 1.

After designing and reviewing my bus shelter in 2D I went on to make a Styrofoam model so I could see it in 3D and look at it from any angle.
I looked at the bad points about it so when making my next model I know what to include or make better.
This model lacks many features I drew in my 2D design such as the lights, windows and curved seating with drainage holes.
I also have to make the bench slightly longer in the middle so it can touch both sides. The final bad point would be in the design, after looking at it and thinking about how some one would sit and their back posture I realised that because the whole back is curved that this will affect how someone sits, to correct this in my next model for the height someone takes up sitting down I will make it vertical up to a certain point then continue with the curved design.
My next model will feature these components and it will make a more complete model which I will then evaluate.

Figure 4.8 Part of a student's development of a product using modelling to test ideas.

To be successful you will:

C. Design and development – **Develop**

Level of response	Mark range
Produce a final design proposal through development that is significantly different and improved compared with any previous alternative design ideas. **(1 mark)** Present a final design proposal that includes technical details of materials and/or components, processes and techniques. **(1 mark)** Produce scale models using traditional materials or 2D and/or 3D computer simulations in order to test important aspects of the final design proposal against relevant design criteria. **(1 mark)** Use client/user group feedback for final modifications. **(1 mark)**	7–10
Develop appropriate designs that use details from alternative design ideas to change, refine and improve the final design proposal. **(1 mark)** Present a final design proposal that includes some details of materials and/or components, processes and techniques. **(1 mark)** Use modelling using traditional materials to test some aspects of the final design proposal against relevant design criteria. **(1 mark)**	4–6
Present development from alternative design ideas that are minor and cosmetic. **(1 mark)** Present a final design proposal that includes superficial details of materials and/or components, processes and techniques. **(1 mark)** Use simple models to test an aspect of the final design proposal against a design criterion. **(1 mark)**	1–3

Communicate (6 marks)

When presenting your design and development work, it is essential that your ideas are communicated effectively. This can be achieved in the following ways.

Through design and development work

You should show evidence of 'design thinking' using any form of effective communication that you feel is appropriate. However, you should try to use a range of skills that may include freehand sketching in 2D and 3D, cut and paste techniques and the use of ICT. It is important to demonstrate a high degree of graphical skill, which will be shown through the accuracy and precision of your work.

When using ICT, you should ensure that it is used appropriately rather than simply for show. For example, specialist CAD software to produce 3D rendered images is likely to be more appropriately used as part of development or final presentation, rather than for initial ideas.

Through presentation graphics and technical drawings

To effectively communicate final designs, a range of skills and drawing techniques should be demonstrated, which could include:

- **pictorial drawings**– isometric, planometric (axonometric), oblique and perspective drawings to convey a 3D representation of the product

- **working drawings** –1st or 3rd angle orthographic, exploded assembly and sectional drawings to convey technical information

- **computer generated** – pictorial and working drawings, renderings, etc. using specialist software.

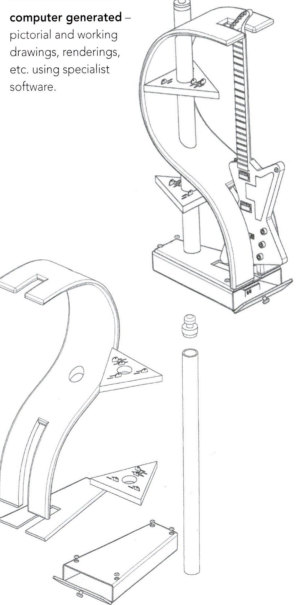

Figure 4.9 A student's CAD drawings for a point-of-sale display for a guitar.

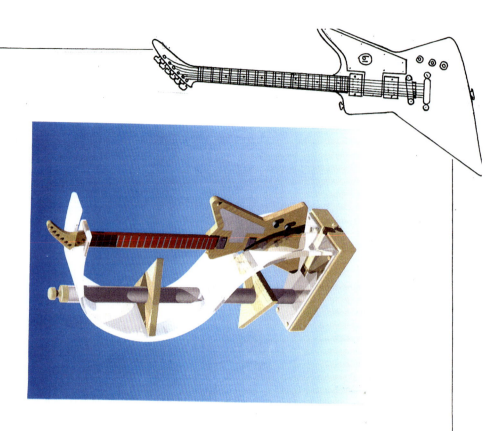

PRESENTATION VISUAL

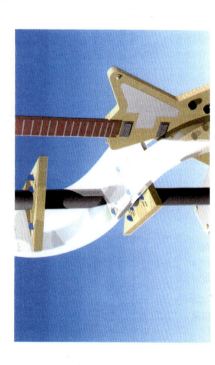

Figure 4.9 continued…

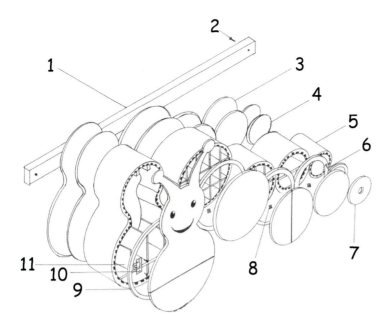

Part no.	Description
1	Baton back support
2	screw
3	7mm MDF base board
4	10mm MDF circles
5	5mm flexi–MDF
6	Chain
7	7mm MDF doors
8	Velcro
9	10mm MDF rings
10	Hinges
11	Hard board shelves

Figure 4.10 *A student's hand-drawn exploded assembly drawing of an educational pack for a nature trail shows the various components and quantities in order to manufacture the product.*

Through the quality of written communication

Annotation should be used to explain design details and convey technical information. You should make sure that the information is presented in a logical order that is easily understood. Specialist technical vocabulary should be used consistently and with precision. Information presented in this section should enable your design thinking and manufacturing intentions to be clearly understood by others and allow third-party manufacture of the final design proposal.

To be successful you will:

C. Design and development – **Communicate**

Level of response	Mark range
Use a range of communication techniques and media, including ICT and computer-aided design (CAD), **(1 mark)** with precision and accuracy **(1 mark)** to convey enough detailed and comprehensive information to enable a third-party to manufacture the final design proposal. **(1 mark)**	4–6
Use a range of communication techniques, including ICT, **(1 mark)** with sufficient skill **(1 mark)** to convey an understanding of design and develop intentions and construction details of the final design proposal. **(1 mark)**	1–3

D Planning (6 marks)

In this section, you will produce a detailed production plan that explains the sequence of operations carried out during the commercial manufacture of your product. It should feature appropriate commercial practices and focus closely upon the identified scale of production. Therefore, workshop practices for the production of your scale model or working prototype should not be described.

You should produce a work order or schedule that illustrates the sequence of operations used during commercial manufacturing. This could be evidenced in the form of a flow chart. The work order should include the order of assembly of parts and components, featuring the necessary tools, equipment and processes to be used during manufacture in volumes higher than 'one-off' production (unless the designed product is specifically a one-off item).

An important part of planning is the use of time, so you must ensure that you consider realistic timescales and deadlines. Where Gantt or time charts are used, you must ensure that they are detailed, cover all aspects of manufacture and include achievable deadlines.

You should identify QC points throughout the product's manufacture and describe all relevant quality checks used. This could be presented as part of a flow chart. Safety checks should also be included as part of planning.

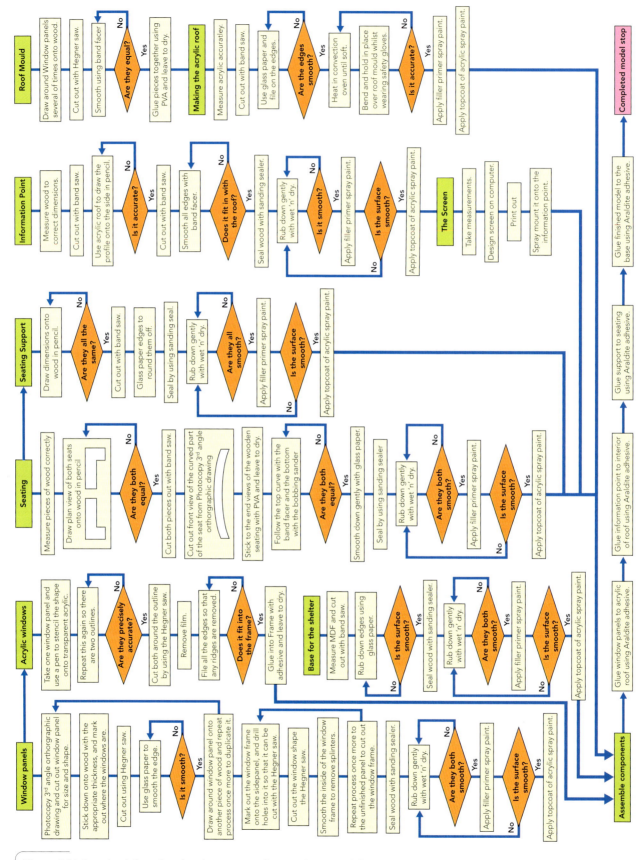

Figure 4.11 *A student's flow chart for planning manufacture of a 3D model.*

To be successful you will:

D. Planning

Level of response	Mark range
Produce a detailed production plan that considers the main stages of manufacture in the correct sequence appropriate to the scale of production. **(1 mark)** Evidence realistic and achievable timescales and deadlines for the scale of production. **(1 mark)** Show quality and safety checks that are justified. **(1 mark)**	4–6
Produce a production plan that considers the main stages of manufacture. **(1 mark)** Show reference to time and scale of production. **(1 mark)** Evidence superficial quality and safety checks. **(1 mark)**	1–3

E Making

FACTFILE:

- Making carries the most marks but is sub-divided into three areas: use of tools and equipment, quality, and complexity/level of demand, making it easier for you to access the marks.
- It is important that all stages of the manufacturing process are photographed to evidence that the product is complete, expertly made, well finished, etc.
- You must ensure that photographs clearly show any details of advanced skills, technical content, levels of difficulty and complexity of construction, so that you can achieve the marks you deserve.
- It is unlikely that a single photograph will be enough to communicate all of the information required, so it will be better to take a series of photographs over a period of time during making.

Use of tools and equipment (9 marks)

You should demonstrate your ability to use tools and equipment with high levels of skill and accuracy and to select appropriate tools and equipment for specific purposes. It is important that you use a range of tools and equipment that allows you to fully demonstrate your skills.

Where computer-aided manufacture (CAM) is a feature of your work, you should make sure that there is plenty of opportunity within the product's manufacture to demonstrate other skills and competencies that you have acquired. For example, do not over-use CAM – make sure that you make the majority of your product/model using hand tools and the appropriate machinery.

You should also work safely and be fully aware of the risks involved when using tools and equipment and the precautions that should be taken to minimise those risks. Appropriate risk assessments for major practical activities could be recorded. Alternatively, stages where health and safety is important could be highlighted and explained within your photographic evidence of the manufacturing process.

LINKS TO:

Unit 2: Health and Safety and **Unit 1: Making** for the procedures for carrying out a risk assessment according to the Health and Safety Executive (HSE).

To be successful you will:

E. Making – Use of tools and equipment

Level of response	Mark range
Select tools and equipment for specific uses independently. **(1 mark)** Use tools and equipment with precision and accuracy. **(1 mark)** Show a high level of safety awareness, for self and others, when using specific tools and equipment. **(1 mark)**	7–9
Select appropriate tools and equipment with some guidance. **(1 mark)** Use tools and equipment with some skill and attention to detail. **(1 mark)** Show sufficient levels of safety awareness, for self and others, when using specific tools and equipment. **(1 mark)**	4–6
Select general tools and equipment with guidance. **(1 mark)** Use tools and equipment with limited skill and attention to detail. **(1 mark)** Show a limited level of safety awareness, for self and others, when using specific tools and equipment. **(1 mark)**	1–3

Model of theme park with vacuum-formed components

Vacuum-formed components continued . . . .

Selecting tools and processes

Due the success of vacuum forming the four pods (instead of turning two and sawing in half), I decided to use the vacuum former again for the sails.

After shaping the MDF mould, I used the vacuum former to produce FOUR identical copies of the sail components. It would have taken a lot more effort to produce these individually and I probably wouldn't have produced them all to the same standard.

When using the vacuum forming machine I was extremely conscious of the heat it generated in order to soften the polystyrene sheet. Therefore, I was careful not to touch any hot surfaces.

The next hazard was the risk of trapping fingers in the many movable parts and clamping mechanisms. I made sure that I was paying full attention throughout the whole process.

Lastly, to prevent the plastic sheet from melting and burning I carefully timed the process and ensured that the machine was under my supervision for the duration.

Health & Safety

Figure 4.12 *Part of a student's documentation for selecting tools and identifying health and safety issues when making.*

Quality (16 marks)

During the manufacture of your product, you should demonstrate your understanding of a range of materials and their working properties. You should be able to select and justify the use of materials that are appropriate to the needs of the product and that match the requirements of your product specification. When selecting materials, you should be able to justify your choice by referring to material properties and suitability for their intended use. The selection and use of appropriate processes and techniques should enable you to produce a high-quality final product that fully matches the final design proposal in all respects.

It is important that all stages of the manufacturing process are photographed to evidence that the product is complete, expertly made, well finished, fully functioning, etc. For this reason you must ensure that photographs clearly show any details of advanced skills, technical content, levels of difficulty and complexity of construction, so that you can achieve the marks you deserve.

It is unlikely that a single photograph will be enough to communicate all of the information required, so it will be better to take a series of photographs over a period of time during making.

ACTIVITY:

At each stage in the manufacture of your product or model ensure that you have photographs of the components you are making. You could use a digital camera or a mobile phone that could then be easily downloaded onto your computer at school or at home to construct your pages.

Treat this activity like a diary of making where you write up what you have achieved each lesson, problems encountered and explanations of why you tackled tasks in the manner you decided.

DO NOT LEAVE THIS SECTION UNTIL THE END – you will have far more work to catch-up on and you may well fail to document some important stages in the manufacture process.

To be successful you will:

E. Making – Quality

Level of response	Mark range
Display a detailed understanding of the working properties of materials used (**1 mark**) with justification for your selection. (**1 mark**) Display a justified understanding of the use of manufacturing processes. (**1 mark**) Produce a high-quality product (**1 mark**) that matches all aspects of the final design proposal (**1 mark**) and is fully functional. (**1 mark**)	11–16
Display a good understanding of the working properties of materials used (**1 mark**) with relevant reasons for your selection. (**1 mark**) Display a good understanding of the use of relevant manufacturing processes. (**1 mark**) Produce a product that matches the final design proposal (**1 mark**) and functions adequately. (**1 mark**)	6–10
Display a limited understanding of the working properties of materials used (**1 mark**) with limited reasoning for your selection. (**1 mark**) Display a limited understanding of the use of manufacturing processes. (**1 mark**) Produce a product that barely matches the final design proposal (**1 mark**) and functions poorly. (**1 mark**)	1–5

Complexity/level of demand (9 marks)

It is important that you demonstrate demanding and high-level making skills in order to achieve high marks. For this reason it is very important that the manufacture of your product offers enough complexity and challenge to gain the maximum credit possible.

The level of complexity of the intended product will already have been established through the finalisation of the design proposal, so it is important that you consider this at an early stage to maximise your potential when manufacturing the product.

You should try to set challenges and demands appropriate to your skill levels and beyond, so that you do not work within your 'comfort zone' and fail to achieve what you are actually capable of.

You should avoid producing simplistic and undemanding work that, however well it is manufactured using appropriate tools, equipment and processes, is unchallenging. This approach cannot result in high levels of credit.

Making of The Bus Shelter

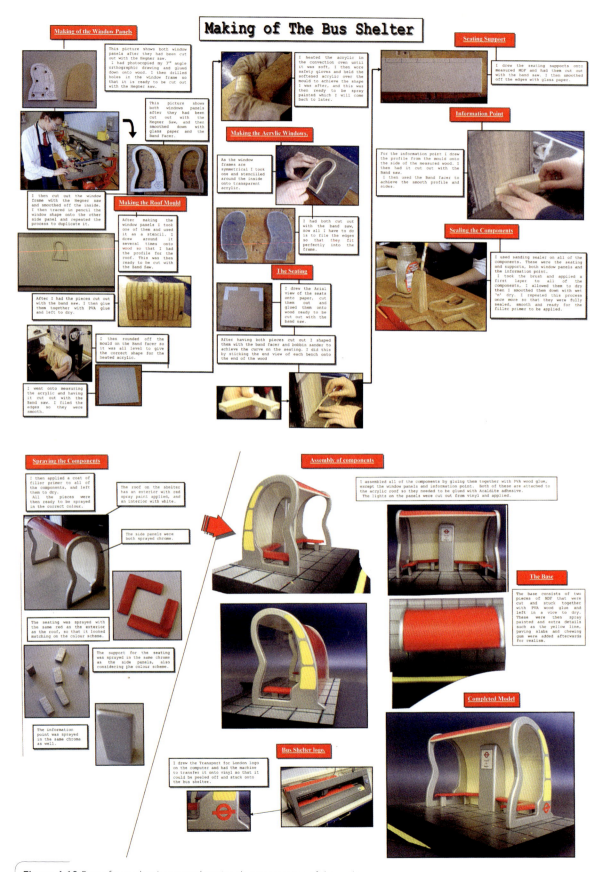

Figure 4.13 *Part of a student's comprehensive documentation of the making process.*

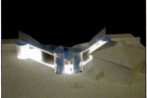

Figure 4.14 *Examples of a student's architectural models requiring a range of modelling skills and attention to detail.*

To be successful you will:

E. Making – Complexity/level of demand

Level of response	Mark range
Undertake a complex and challenging making task. (1 mark) Include a wide range of skills, (1 mark) demonstrating precision and accuracy in their use. (1 mark)	7–9
Undertake a reasonably complex making task that offers some challenge. (1 mark) Include a range of skills, (1 mark) demonstrating attention to detail in their use. (1 mark)	4–6
Undertake a simple and undemanding making task. (1 mark) Include a limited range of skills, (1 mark) requiring little attention to detail in their use. (1 mark)	1–3

F Testing and evaluating
(10 marks)

Once you have completed the manufacture of your product, you should carry out tests to check its fitness for purpose with reference to commercial techniques where possible.

Your finished product should be tested under realistic conditions, wherever possible, to decide on its success using the points of specification to check the product's performance and quality. You should describe in detail any tests carried out and justify them by stating what is being tested and why. Tests should be objective and many should be carried out by the client/user group. In addition, the involvement of other potential users would be a reliable way of gathering unbiased and reliable third-party feedback.

Well-annotated photographic evidence is a good tool to use when describing testing. You should use the results of your testing and views of the client/user group to help evaluate your final product.

Your evaluation should relate to the measurable points of your product specification and should be as objective as possible. You should use the information from your testing, evaluation and client/user group feedback to make suggestions for possible modifications and future improvements to the product. Suggestions for modifications should focus on improving the performance of the product or its quality, and not simply cosmetic changes.

ACTIVITY:

Many students often overlook the testing and evaluation of a final product as 'not that important'. However, it is extremely important in wrapping up your project and bringing it to a logical conclusion. In order to carry out effective testing and evaluation of your final product you need to consider the following:

- Testing against your initial design specification to determine whether it satisfies the criteria. This will involve the justification of each specification point.
- Gaining third-party feedback from your client and/or user group using appropriate questionnaires or interview prompts.
- Writing an objective evaluation that discusses both the positive and negative aspects of your project.
- Suggesting modifications as a result of your evaluation. Remember – no design is ever perfect so there are bound to be aspects that need to be improved if you had more time. More importantly, there are bound to be technical problems that you did not have the time to address in detail that would need further development.

Finally, you should check the sustainability of your final product by carrying out a life-cycle assessment (LCA) to assess its impact on the environment. The most important life-cycle stages to consider when carrying out an LCA of your final product are below.

- **Raw materials** – What impact does the extraction of the raw materials in your product have upon the environment? Could you use fewer materials? Could you use recyclable materials?

- **Manufacture** – Are the processes you have identified energy efficient? Can manufacturing and assembly processes be simplified? How can the amount of waste produced be minimised?

- **Distribution** – How can transportation mileage be minimised? Can the design be simplified so as to reduce or lighten materials?

- **Use** – Will your product last a long time and can it be repaired if something goes wrong? How can you promote its efficient use? Can you use its green credentials to positively market your product?

- **End-of-life** – Can your product be recycled or reused? How can you reduce the amount of waste it produces from ending up as landfill?

LINKS TO:

Unit 3: Designing for the Future: Sustainability looks at these issues in greater detail.

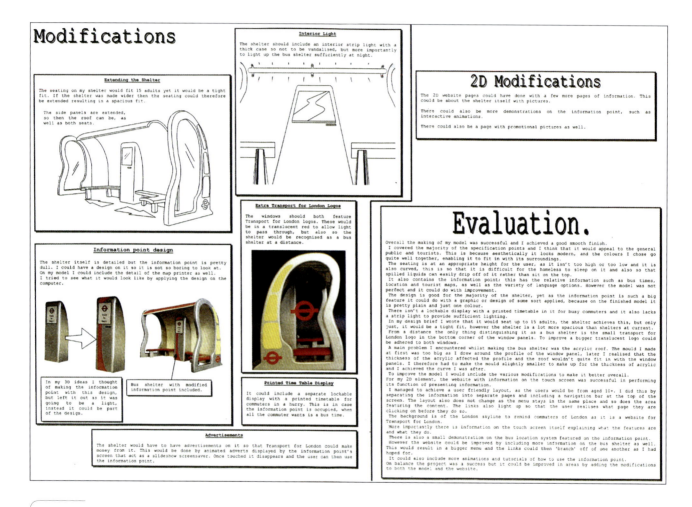

Figure 4.15 No design is ever perfect – suggestions for further improvement are always useful.

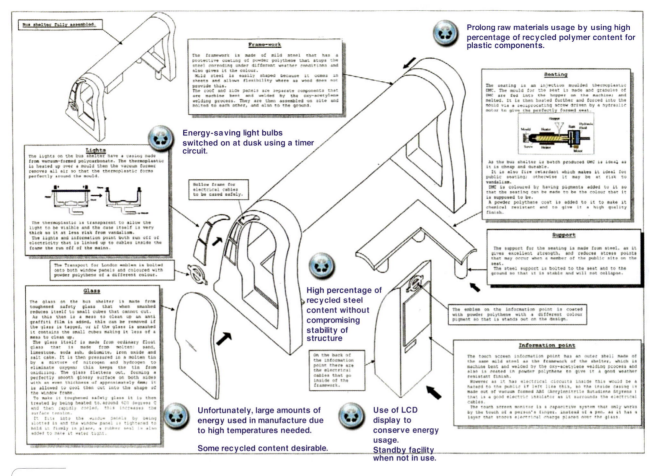

Figure 4.16 *Environmental issues relating to the final design are highlighted.*

To be successful you will:

F. Testing and evaluating

Level of response	Mark range
Carry out a range of tests that are justified in order to check the performance and/or quality of the final product. **(1 mark)** Objectively evaluate, including third-party evaluation, considering most relevant, measurable specification points in detail. **(1 mark)** Suggest modifications that are justified from tests carried out; focus on improving performance and/or quality of the final product. **(1 mark)** Carry out relevant and useful LCA on the final product to check its sustainability. **(1 mark)**	7–10
Carry out a range of tests to check the performance and/or quality of the final product. **(1 mark)** Use objective evaluation with reference to most specification points. **(1 mark)** Suggest relevant modifications that are justified from tests that were carried out. **(1 mark)**	4–6
Carry out one or more simple tests to check the performance and/or quality of the final product. **(1 mark)** Use subjective and superficial evaluation with reference to a few specification points. **(1 mark)** Suggest only simple cosmetic modifications. **(1 mark)**	1–3

Index